Rethinking the Human in the Darwinian Novel
Zola, Hardy, and Utopian Fiction

LEGENDA

LEGENDA is the Modern Humanities Research Association's book imprint for new research in the Humanities. Founded in 1995 by Malcolm Bowie and others within the University of Oxford, Legenda has always been a collaborative publishing enterprise, directly governed by scholars. The Modern Humanities Research Association (MHRA) joined this collaboration in 1998, became half-owner in 2004, in partnership with Maney Publishing and then Routledge, and has since 2016 been sole owner. Titles range from medieval texts to contemporary cinema and form a widely comparative view of the modern humanities, including works on Arabic, Catalan, English, French, German, Greek, Italian, Portuguese, Russian, Spanish, and Yiddish literature. Editorial boards and committees of more than 60 leading academic specialists work in collaboration with bodies such as the Society for French Studies, the British Comparative Literature Association and the Association of Hispanists of Great Britain & Ireland.

The MHRA encourages and promotes advanced study and research in the field of the modern humanities, especially modern European languages and literature, including English, and also cinema. It aims to break down the barriers between scholars working in different disciplines and to maintain the unity of humanistic scholarship. The Association fulfils this purpose through the publication of journals, bibliographies, monographs, critical editions, and the MHRA Style Guide, and by making grants in support of research. Membership is open to all who work in the Humanities, whether independent or in a University post, and the participation of younger colleagues entering the field is especially welcomed.

ALSO PUBLISHED BY THE ASSOCIATION

Critical Texts
Tudor and Stuart Translations • *New Translations* • *European Translations*
MHRA Library of Medieval Welsh Literature

MHRA Bibliographies
Publications of the Modern Humanities Research Association

The Annual Bibliography of English Language & Literature
Austrian Studies
Modern Language Review
Portuguese Studies
The Slavonic and East European Review
Working Papers in the Humanities
The Yearbook of English Studies

www.mhra.org.uk
www.legendabooks.com

STUDIES IN COMPARATIVE LITERATURE

Studies in Comparative Literature are produced in close collaboration with the British Comparative Literature Association, and range widely across comparative and theoretical topics in literary and translation studies, accommodating research at the interface between different artistic media and between the humanities and the sciences.

ALSO PUBLISHED IN THIS SERIES

20. *Aestheticism and the Philosophy of Death: Walter Pater and Post-Hegelianism*, by Giles Whiteley
21. *Blake, Lavater and Physiognomy*, by Sibylle Erle
22. *Rethinking the Concept of the Grotesque: Crashaw, Baudelaire, Magritte*, by Shun-Liang Chao
23. *The Art of Comparison: How Novels and Critics Compare*, by Catherine Brown
24. *Borges and Joyce: An Infinite Conversation*, by Patricia Novillo-Corvalán
25. *Prometheus in the Nineteenth Century: From Myth to Symbol*, by Caroline Corbeau-Parsons
26. *Architecture, Travellers and Writers: Constructing Histories of Perception*, by Anne Hultzsch
27. *Comparative Literature in Britain: National Identities, Transnational Dynamics 1800-2000*, by Joep Leerssen
28. *The Realist Author and Sympathetic Imagination*, by Sotirios Paraschas
29. *Iris Murdoch and Elias Canetti: Intellectual Allies*, by Elaine Morley
30. *Likenesses: Translation, Illustration, Interpretation*, by Matthew Reynolds
31. *Exile and Nomadism in French and Hispanic Women's Writing*, by Kate Averis
32. *Samuel Butler against the Professionals: Rethinking Lamarckism 1860–1900*, by David Gillott
33. *Byron, Shelley, and Goethe's Faust: An Epic Connection*, by Ben Hewitt
34. *Leopardi and Shelley: Discovery, Translation and Reception*, by Daniela Cerimonia
35. *Oscar Wilde and the Simulacrum: The Truth of Masks*, by Giles Whiteley
36. *The Modern Culture of Reginald Farrer: Landscape, Literature and Buddhism*, by Michael Charlesworth
37. *Translating Myth*, edited by Ben Pestell, Pietra Palazzolo and Leon Burnett
38. *Encounters with Albion: Britain and the British in Texts by Jewish Refugees from Nazism*, by Anthony Grenville
39. *The Rhetoric of Exile: Duress and the Imagining of Force*, by Vladimir Zorić
40. *From Puppet to Cyborg: Pinocchio's Posthuman Journey*, by Georgia Panteli
41. *Utopian Identities: A Cognitive Approach to Literary Competitions*, by Clementina Osti
43. *Sublime Conclusions: Last Man Narratives from Apocalypse to Death of God*, by Robert K. Weninger
44. *Arthur Symons: Poet, Critic, Vagabond*, edited by Elisa Bizzotto and Stefano Evangelista
45. *Scenographies of Perception: Sensuousness in Hegel, Novalis, Rilke, and Proust*, by Christian Jany
46. *Reflections in the Library: Selected Literary Essays 1926–1944*, by Antal Szerb
47. *Depicting the Divine: Mikhail Bulgakov and Thomas Mann*, by Olga G. Voronina
48. *Samuel Butler and the Science of the Mind: Evolution, Heredity and Unconscious Memory*, by Cristiano Turbil
49. *Death Sentences: Literature and State Killing*, edited by Birte Christ and Ève Morisi
50. *Words Like Fire: Prophecy and Apocalypse in Apollinaire, Marinetti and Pound*, by James P. Leveque

Rethinking the Human in the Darwinian Novel

Zola, Hardy, and Utopian Fiction

❖

Niall Sreenan

LEGENDA

Studies in Comparative Literature 51
Modern Humanities Research Association
2025

Published by Legenda
an imprint of the Modern Humanities Research Association
Salisbury House, Station Road, Cambridge CB1 2LA

ISBN 978-1-78188-847-6

First published 2025

Copy-Editor: Charlotte Wathey

CONTENTS

ACKNOWLEDGEMENTS

This book's genesis began in 2017, shortly after the conclusion of my PhD, during which time it survived three house moves, the birth of two children, a global pandemic, and the vicissitudes of early academic career under- and over-employment. That the book exists at all is a tribute to a multiplicity of supportive friends, colleagues, and interlocutors, as well as to the inspiration and incredible forbearance of my family. I will try to mention by name as many of them as I can below.

Firstly, I would like to acknowledge the support and influence of my two doctoral supervisors, Tim Mathews and Elinor Shaffer, to whom I will always be indebted for taking me on as a slightly unlikely PhD candidate. Tim's intellectual generosity and critical acuity have been instrumental in the development of the book's argument; I continue to be grateful for his influence and support. I wish also to thank Elinor, from whose depth of knowledge on Darwin, reception, and the interface of literature and science, I have benefitted enormously. Many others at UCL including (but certainly not limited to) Florian Mussgnug, Stephanie Bird, and Tim Beasley Murray, offered incisive comments and thoughts in contexts formal and informal that helped shape the project, as did the support and comradeship of my friends and peers, Mathelinda Nabugodi, Liz Harvey, and Stefano Rossoni. David Amigoni's encouragement to pursue this project further is also greatly appreciated.

I am grateful to the editorial board of Studies in Comparative Literature for agreeing to publish this book and I would like to single out Graham Nelson at Legenda, to whom I owe sincere thanks. His patience kept the book alive, even when its heartbeat was dimmed. Earlier versions of some of the arguments made in this book appeared in *Island Studies Journal*, *Modern & Contemporary France* and the proceedings of conferences run by the Émile Zola Society of London in 2013 and the Hermes Consortium for Literary and Cultural Studies in 2015; I am grateful to the editors of these publications for offering my work a forum.

I have also been lucky to have had career support and advice from many colleagues and peers both within and without the academic world: in particular, thank you to Rosa Mucignat, Ziad Elmarsafy, Steen Bille Jorgensen (and others from the *Landscapes of Realism* project), as well as Benedict Wilkinson for references written, invitations extended, and tips offered. A big thank you should also go to my colleagues and friends in the School of Modern Languages at the University of St Andrews, who have put enormous trust in me as a new (and, again, slightly unlikely) colleague — a trust I hope to reward with more books in future.

Lastly, I owe my family the most profound thanks of all: to the backing of my parents and brother; to my two sons, Arlo and Odhran, whose wit, exuberance for life, and beauty is an inspiration; and to my wife, Claire, whose forbearance, pragmatism, encouragement, fortitude, wisdom, and love this book owes its existence, and to whom it is dedicated.

N.S., August 2025

ABBREVIATIONS

The following abbreviations are used for subsequent, in-text references to the primary works under consideration:

Origin:	Charles Darwin, *On the Origin of Species by Means of Natural Selection, or the Preservation of Favoured Races in the Struggle for Life*, 2nd edn (John Murray, 1860)
Descent:	Charles Darwin, *The Descent of Man, and Selection in Relation to Sex*, 1st edn, 2 vols (John Murray, 1871)
Germinal:	Émile Zola, *Germinal*, ed. by Olivier Lumbroso and Henri Mitterand, in *Œuvres complètes. Tome 12: Souffrance et révolte (1884–1885)* (Nouveau Monde, 2005); *Germinal*, trans. by Peter Collier (Oxford University Press, 2008)
L'Œuvre/Masterpiece:	Zola, Émile, *L'Œuvre*, ed. by Henri Mitterand and Olivier Got, in *Œuvres complètes. Tome 13: Naturalisme pas mort! (1886–1888)* (Nouveau Monde, 2005); *The Masterpiece*, trans. by Roger Pearson and Thomas Walton, rev. edn (Oxford University Press, 2008)
Le Ventre/Belly:	Émile Zola, *Le Ventre de Paris*, ed. by Henri Mitterand, Patricia Carles, and Béatrice Desgranges, *Œuvres Complètes. Tome 5: Thiers au pouvoir (1871–1873)* (Nouveau Monde, 2005); *The Belly of Paris*, trans. by Brian Nelson (Oxford University Press, 2009)
Blue Eyes:	Thomas Hardy, *A Pair of Blue Eyes*, ed. by Alan Manford (Oxford University Press, 2005)
Return:	Thomas Hardy, *The Return of the Native*, ed. by Simon Avery (Broadview Press, 2013)
Erewhon:	Samuel Butler, *Erewhon*, ed. by Peter Mudford (Penguin, 1970)
Island:	Aldous Huxley, *Island*, ed. by David Bradshaw (Vintage Books, 2005)
Possibilité/Possibility:	Michel Houellebecq, *La Possibilité d'une île* [2005] (Fayard/ Livre de Poche, 2007); *The Possibility of an Island*, trans. by Gavin Bowd (Vintage, 2007)

INTRODUCTION

❖

Rethinking the Human after Darwin

> Bien sûr, c'est à la science que doivent s'adresser les romanciers et les poètes, elle est aujourd'hui l'unique source possible. Mais, voilà! que lui prendre, comment marcher avec elle? Tout de suite, je sens que je patauge. — ÉMILE ZOLA[1]
>
> [Science, of course, is what poets and novelists are going to have to turn to; science is their only possible source these days. But there you are again! What are they to get out of it? How are they to keep up with it? As soon as I think of that, my mind goes blank.]

> In the course of centuries the naïve self-love of men has had to submit to two major blows at the hands of science. The first was when they learnt that our earth was not the centre of the universe but only a tiny fragment of a cosmic system of scarcely imaginable vastness. This is associated in our minds with the name of Copernicus [...]. The second blow fell when biological research destroyed man's supposedly privileged place in creation and proved his descent from the animal kingdom and his ineradicable animal nature. This revaluation has been accomplished in our own days by Darwin, Wallace and their predecessors, though not without the most violent contemporary opposition. — SIGMUND FREUD[2]

Émile Zola's *L'Œuvre* (1886), the fourteenth work in his twenty-novel cycle, *Les Rougon-Macquart*, posits a radical scientifico-aesthetic programme. '[L]a science', argues the fictional novelist, Sandoz, is the only possible foundation for an adequately post-Romantic realist literary aesthetic. Subtitled as *Histoire naturelle et sociale d'une famille sous le Second Empire*, Zola's sprawling literary-scientific project draws on the sciences of heredity, experimental medicine, and positivist social theory to realise in social and biological totality the history of a single fictional family across the span of Second Empire France. Positioning his work in opposition to the individual social scope of Honoré de Balzac's 'zoologie humaine', Zola frames

1 Émile Zola, *L'Œuvre*, ed. by Henri Mitterand and Olivier Got, in *Œuvres complètes. Tome 13: Naturalisme pas mort! (1886–1888)* (Nouveau Monde, 2005), p. 44 (*hereafter referenced as L'Œuvre*); *The Masterpiece*, trans. by Roger Pearson and Thomas Walton, rev. edn (Oxford University Press, 2008), pp. 37–38 (hereafter referenced as *Masterpiece*).

2 Sigmund Freud, *Introductory Lectures on Psycho-Analysis (1916–1917)*, in *The Standard Edition of the Complete Psychological Works of Sigmund Freud*, ed. and trans. by James Strachey and others, 24 vols (Hogarth Press/ Institute of Psycho-Analysis, 1953–74), XVI (1963), p. 284.

his task as that of the naturalist and the physiologist, focusing not on individuals, but on the mechanisms and laws of environment and nature that constitute them.[3] The fictional Sandoz has in mind a similarly transcendent literary scientific project: the creation of a series of works that move beyond philosophy and academic distinctions into a work 'où l'on tâcherait de mettre les choses, les bêtes, les hommes [one would endeavour to put everything, the beast of the field as well as mankind] (*L'Œuvre*: 44; *Masterpiece*: 37). Zola and his fictional proxy, therefore, both seek to reconceptualise and reframe literature's capacity for the apprehension of social and biological reality by a widening of realist literature's representational and epistemological aperture. It is a vision that includes not merely the human species as family but the social and biological forces that determine human as well as non-human life.

Through Sandoz, Zola's work raises a series of fundamental questions as to the precise methodological and philosophical coordinates of this literary-scientific fantasy. To what kind of science does one turn to capture social, historical, and organic reality in all its dynamism, 'en plein coulée de la vie universelle' [in the mighty universal flow of life] (*L'Œuvre*: 44; *Masterpiece*: 37)? And how does one account for the similarly dynamic nature of scientific knowledge production, both too large in scope and too mutable in outlook for the novelist easily 'marcher avec elle' [to keep up with it] (*L'Œuvre*: 44; *Masterpiece*: 38)? Indeed, by thematising this ambition in his own work, Zola draws attention to its textuality, implying further questions about the capacity of the written word to break the boundaries of its own symbolic containment. To what extent can writing, whether scientific or literary, reach beyond its own discursive limitations to apprehend reality?

For Sigmund Freud, writing nearly thirty years later in the early part of the twentieth century, the march of scientific progress promised a wholesale transformation in our understanding of humanity. Yet in his *Vorlesungen zur Einführung in die Psychoanalyse* [*Introductory Lectures on Psycho-Analysis*] (1916–17), scientific advancement is understood to challenge the very vaulting epistemological and representational ambition exemplified by Zolian Naturalism. Freud argues that a series of Copernican revolutions in science since the sixteenth century until his own day represent a series of incursions or 'major blows' to the human ego and in turn on its grasp on reality.[4] Specifically, three instances of what Thomas Kuhn would later characterise as a scientific 'paradigm shift' displace humanity from its self-appointed central place in the cosmos, in the natural world, and in its own psychic territory: first through the inauguration of a heliocentric conception of the cosmos, then through the emergence of Darwinian evolution, and finally, through the advent of psychoanalysis.[5] Far from offering a more secure basis upon which representational knowledge can be constructed, Freud, Darwin, and Copernicus

3 Émile Zola, 'Différences entre Balzac et moi (1869)', in *Comment Emile Zola composait ses romans: d'après ses notes personnelles et inédites*, ed. by Henri Massé (Charpentier, 1906), pp. 24–26.

4 Freud, *Introductory Lectures on Psycho-Analysis*, p. 284.

5 Similarly, for Kuhn, the most consequential and threatening element of Darwin's 'scientific' revolution was the 'abolition of that teleological kind of evolution' according to which the '"idea" of man' was present at creation and represented the telos of progress: Thomas S. Kuhn, *The Structure of Scientific Revolutions*, 3rd edn (University of Chicago Press, 1970), p. 172.

each challenge what Freud takes care to designate as the coordinates of a specifically 'human megalomania'.[6] It is not only that these revolutionary movements in scientific understanding challenge the possibility of knowing and being as such in a psychologically and biologically decentred cosmos. By specifying the human quality of such delusions, Freud underlines the symbiotic relationship between the very concept of the human and pre-Copernican, pre-Darwinian, and pre-Freudian fantasies of transcendence and supremacy. Contra Zola and Sandoz, certain kinds of scientific knowledge threaten a specifically human narcissism that is constitutive both of the realist fantasy of totalising representation as well as of the human as both subject and species, both of which rely on a position of epistemic and ontological exception. At the core of Freud's argument is psychic resistance: we know Freudian psychoanalysis, Darwinism, and Copernican heliocentrism are truly revolutionary, Freud argues, because they elicited (and continue to elicit) widespread opposition. Resistance to Darwinism is not a function of its conceptual difficulty or abstraction: 'Its source,' Freud says, 'lies deeper'; what is at stake is a fundamental fantasy, a belief in human as the central axis of cosmological, biological, and psychical existence.

A near contemporary of both Zola and Freud, the Victorian novelist Thomas Hardy asked similar questions about the relation of the project of literary realism to scientific representation and thought. Writing in an essay, 'The Science of Fiction', Hardy observes that Zola's 'work on the Roman Experimental [*sic*], seems to reveal an obtuseness to the disproof conveyed in his own novels'.[7]Attempts at scientifically realist literature may be 'admirable', Hardy admits, but ignore the subjectivity, selectiveness, and affective charge involved in the creation and reception of literature, which are inimical to an objective and totalising view of a notionally scientific form of realist representation. Indeed, he argues, the notion of a scientifically realist literature, a literature of 'copyism', over-estimates the objectivity and finality of scientific knowledge itself. Yet, despite his antipathy to the Zolian project, Hardy seeks to articulate a positive relation between science and literature. 'Art,' he writes, 'is science with an addition': literary art is neither reducible to what he designates as the scientific pursuit of 'comprehensive and accurate knowledge of realities', nor is it opposed to scientific practice and discourse.[8] Literary discourse supplements the notionally complete vision of the world thought to be offered by science. We might understand Hardy's argument here in proto-deconstructive terms: literature's relationship to science is defined by a relation of what Jacques Derrida calls 'supplementarity': 'a surplus, a plenitude enriching another plenitude'.[9] By this logic, the 'constructive' act of story-telling, which by Hardy's analysis is the condition of possibility of science, is also an act of

6 Ibid, p. 285.

7 Thomas Hardy, 'The Science of Fiction', *The New Review*, 4 (1891), pp. 315–19; repr. in *Thomas Hardy's Public Voice: The Essays, Speeches, and Miscellaneous Prose*, ed. by Michael Millgate (Clarendon Press, 2001), pp. 106–10.

8 Ibid., p. 106.

9 Jacques Derrida, *Of Grammatology*, trans. by Gayatri Chakravorty Spivak (Johns Hopkins University Press, 1998), p. 144.

negation: fiction's plenitude threatens to undermine the fullness of scientific truth, revealing it as constitutively incomplete.

Freud's reading of scientific progress and Hardy's conception of scientific literature, together, raise further questions for Zola and Sandoz, and for our own understanding of the relation of literary writing (not only in its realist modality) to scientific knowledge. If cosmology, biology, and psychology (in its psychoanalytic form) work to negate human knowledge and epistemic grounded-ness, in what way can the self-consciously scientific novel register this destabilisation? Indeed, if literature and writing is always subjective, selective, and affective, in what way can we understand the relation of literature to its scientific sources? And what position might be found for the human in the relationship of literature to science, when Freud's three Copernican revolutions and Hardy's critique of literary scientism seem to mitigate against the notion of objectivity as such, as well as the very possibility of the human as a transcendent, individual, and self-knowing subject?

This book focuses on Charles Darwin's theory of evolution, the second scientific revolution in Freud's three-act history of intellectual assaults on the human ego. It does so by examining the way in which literature — written in the historical aftermath of the Darwinian revolution — can be read as a response to that assault. How does literature of the post-Darwinian period respond to and register the displacement of the human from its position at the top of a natural order? To what extent can we see literary responses to Darwin as supplementary discourses, offering their own theory of the human, which answers to, integrates, or repudiates Darwin's destruction of a pre-Copernican anthropocentrism?

To answer these questions, this book's critical itinerary is split into three chapters, each one focusing on a different set of novels. The first two chapters are concerned with literary realist responses to Darwin's thought. The first focuses on a constellation of works from Zola's Naturalist series, *Histoire naturelle et sociale d'une famille sous le Second Empire*: *Le Ventre de Paris* (1873) [*The Belly of Paris*], *Germinal* (1885), and the already cited *L'Œuvre* [*The Masterpiece*] (1886).[10] The second chapter examines two works by Thomas Hardy, *A Pair of Blue Eyes* (1873) and *The Return of the Native* (1878).[11] These two chapters examine the way in which the reception of Darwin in French and British nineteenth-century realism registers and interrogates the Darwinian displacement of the human and its attendant epistemological and representational implications. The third and final chapter shifts generic and historical registers, tracing a new historical and generic genealogy of the reception of Darwin in a series of Utopian novels from nineteenth-, twentieth-, and twenty-

10 Émile Zola, *Le Ventre de Paris*, ed. by Henri Mitterand, Patricia Carles, and Béatrice Desgranges, in *Œuvres complètes. Tome 5: Thiers au pouvoir (1871–1873)* (Nouveau Monde, 2005) (hereafter referenced as *Le Ventre*); *The Belly of Paris*, trans. by Brian Nelson (Oxford University Press, 2009) (hereafter referenced as *Belly*); *Germinal*, ed. by Olivier Lumbroso and Henri Mitterand, in *Œuvres complètes. Tome 12: Souffrance et révolte (1884–1885)* (Nouveau Monde, 2005); *Germinal*, trans. by Peter Collier (Oxford University Press, 2008) (both hereafter referenced as *Germinal*).

11 Thomas Hardy, *A Pair of Blue Eyes*, ed. by Alan Manford (Oxford University Press, 2005) (hereafter referenced as *Blue Eyes*); *The Return of the Native*, ed. by Simon Avery (Broadview Press, 2013) (hereafter referenced as *Return*).

first-century Britain and France: Samuel Butler's *Erewhon* (1872), Aldous Huxley's *Island* (1962), and Michel Houellebecq's *La Possibilité d'une île* (2005) (*The Possibility of an Island*, 2005).[12] While these works' speculative and satirical forms are notionally opposed to those of the realist novel of the nineteenth century, the questions asked of them are the same. To what extent can these novels' response to Darwin be read specifically in relation to the Freudo-Darwinian negation of the human subject? And how might these novels offer an implicit theory or future image of a post-Darwinian human?

In the course of this critical narrative, this study engages with both historical and theoretical literary inquiry and scholarship on Darwinism and its relationship to literature. Working with historical material, lineages of reception, influence, and translation, this book's structuring logic is founded on the fact of Darwin's influence on all five of the literary authors that feature across its three chapters. As Gillian Beer has argued, as Darwin's ideas have passed into a form of common sense, one could in theory study responses to Darwinism in any novel written in the wake of Darwin's *On the Origin of Species,* first published in 1859. After Darwin, '[e]veryone found themselves living in a Darwinian world in which old assumptions had ceased to be assumptions, could be at best beliefs, or myths, or, at worst, detritus of the past'.[13] Yet, focusing on novels by authors whom we know have read (or appear to have read) Darwin not only helps place historical and generic constraints on the selection of a corpus that would otherwise be unlimited, it also provides us with a set of materials through which to explore the complex nature of what Beer calls the 'question-raising' act of reading and reception.[14] So, while each of this book's chapters are grounded in the organising principle of conscious reception and influence, they also seek to register and examine the complexity of such reception, which as Freud's analysis would suggest may involve more or less conscious forms of resistance. As we shall see, Darwin's writing and thinking play an ambivalent role in the works studied in this book, as Darwinian evolution informs contrary conceptions of human life and relations, sometimes expressed dialogically within the same texts. As well as paying attention to the divergent forms and expressions that Darwin's work takes in these works, we shall also have to widen our conception of 'reading', since it is not always (or ever) scientifically accurate or theoretically faithful renderings of Darwin that produce these kinds of responses, but glitches in transmission, via complex circuits of translation, misunderstanding, and creative interpretations. This organisational tactic, therefore, is one way of approaching the kinds of literary historical and discursive questions with which Zola's Sandoz is so concerned, with specific regard

12 Samuel Butler, *Erewhon*, ed. by Peter Mudford, new & rev. edn (Penguin, 1970) (hereafter referenced as *Erewhon*); Aldous Huxley, *Island*, ed. by David Bradshaw (Vintage Books, 2005) (hereafter referenced as *Island*); Michel Houellebecq, *La Possibilité d'une île* [2005] (Fayard/ Livre de Poche, 2007) (hereafter referenced as *Possibilité*); *The Possibility of an Island*, trans. by Gavin Bowd [2005] (Vintage, 2007) (hereafter referenced as *Possibility*).

13 Gillian Beer, *Darwin's Plots: Evolutionary Narrative in Darwin, George Eliot and Nineteenth-Century Fiction*, 3rd edn (Cambridge University Press, 2009), p. 3.

14 Ibid., p. 4.

for the phenomenon of the self-consciously post-Darwinian novel. In what ways have the novels by Zola, Hardy, and the three Utopian novelists with which this book is concerned integrated Darwinian evolutionary thought?[15] Through what modes of transmission, reception, translation, and engagement does Darwinian thought come to inform their works? And, to draw once more on Freud's thesis on the wider societal reception of Copernican shifts in human thought, what forms of resistance and receptivity, whether conscious or otherwise, seem to characterise their literary response to Darwinian evolution? Each chapter offers an account of the reception of Darwin, partly through historical, extra-literary, and biographical material, and primarily through an analysis of Darwinian concepts, signifiers, images, motifs, and narratives present in the novels with which they are concerned.

If this book's organisation and corpus revolves around a relation of reception, its ultimate objective is critical and theoretical. Rather than only re-examining whether, how, or in what way Darwin has influenced the authors that feature in this book, it also seeks to explore these authors' literary responses to Darwin's work as forms of philosophical or theoretical thinking in themselves, which can be read as replies to questions about the nature of post-Darwinian human. If, as Malcolm Bowie argued more than two decades ago, what was by then an already 'dilapidated partition between "theory" and "fiction"' allows us to read theorists (like Freud) as 'portraitists' of human life that are comparable to the work of somebody like Marcel Proust, then we too might seek to read figures like Zola or Hardy as producing in the novel form a kind of enquiry in the human sciences alongside and in response to the theoretical questions raised by Darwin.[16] For Bowie, Freud, Proust, and Jacques Lacan shared a common interest in the questions and theories of desire. Here, the object of shared theoretical enquiry is 'the human', addressed, elaborated upon, problematised, and theorised in different ways, first by Darwin, and then, in response, by Zola, Hardy, and in three Utopian fictions by Butler, Huxley, and Houellebecq.

To help draw out, clarify, and situate the theoretical implications of these literary responses to Darwin, this book draws on a range of theoretical voices, another set of 'portraitists', in whose work we find variety of concepts of the human (and indeed the post-human), both implicitly and explicitly in dialogue with Darwinian thought or the literary works written in response to that thought. Of particular importance to this study is the work of Gilles Deleuze (both with and without Félix Guattari), which appears in all three of the chapters that follow. Deleuze's work is notable (if not entirely unique) among the efflorescence of twentieth-century critical, philosophical continental thought in terms of the seriousness and creativity with which it addresses and makes use of scientific concepts as part of a wider critical and philosophical project. This book draws on a range of Deleuzean ideas, across all three of its chapters, which include interventions from Deleuze's

15 I use the proper adjectival term *Utopian* to signal these novels' connection with Thomas More's inaugural *Utopia* (1516); other uses of the term *utopia* are more general and generic.

16 Malcolm Bowie, *Freud, Proust and Lacan: Theory as Fiction* (Cambridge University Press, 1987), pp. 5–6.

reading of Darwin in his major philosophical work, *Différence et répétition* [*Difference and Repetition*] (1968), to Elizabeth Grosz's feminist re-reading of Darwin, which takes inspiration from the work of Henri Bergson (via Deleuze), and Deleuze's early essay on the materialist contradictions of desert islands. Common to each of these is an affirmation, in its dual sense of acknowledgement and endorsement, of what might be called a 'post-human ontological displacement of the human', which for Freud was more commonly a source of violent resistance. This book, therefore, does engage with such thinkers and ideas, such as Donna Haraway's concept of the 'cyborg', but stops short of calling itself an explicitly 'post-human' analysis or project. This is not merely a pragmatic decision to limit this book's theoretical scope, but also an attempt to think the post-human through a set of contextually specific historical and literary resources that may anticipate and inform the 'post-human', without relying on the invocation of that particular overdetermined signifier. This book, as a literary critical project, also focuses largely on questions of language, narrative, and signification, the theoretical justification for which is addressed later in this introduction. And as such, it also makes use of feminist, psychoanalytic, deconstructionist, and political theoretical writing. This not only helps to situate and illuminate the way in which these novels address and rethink the biological, political, cultural, and technological coordinates of the post-Darwinian human but reveals the way in which the 'post-human' sometimes conceals a disavowed recommitment to the human.

The coming-together of these two modes of inquiry or understanding — the historical and the theoretical — also suggests the need for some further reflection on the method and function of comparison in this book. The first two chapters focus on a single novelist working in the realist tradition of the nineteenth century in France (Zola) and Britain (Hardy); the third shifts its generic and historical focus to examine a genealogy of novelists writing in the tradition of the Utopian novel, across these same two contexts (Butler and Huxley in Britain, Houellebecq in France). If the first two chapters of this book can be understood as an exploration of literary realist responses to Darwinian thought in the latter half of the nineteenth century in Britain and France, the third chapter uses this period as the starting point of an analysis that privileges Utopian responses to Darwin, looking beyond the immediate historical aftermath of the publication of Darwin's major works, towards twentieth- and twenty-first-century speculative works of evolutionary fiction. As such, this corpus suggests several axes of comparison. It offers generic (utopian and realist), historical (nineteenth, twentieth, and twenty-first centuries), cultural and national (France and Britain), points of comparison, as well as suggesting a comparative analysis of literary and scientific discourses. Each of these points of comparison, at different points in the book, are highlighted, and inform the sections that are engaged explicitly with reception history, influence, and translation. This is one of the ways in which this book is distinct from some other approaches to Darwin and literature, particularly in anglophone criticism (which will be addressed in more detail later in this introductory chapter): the work of path-breaking critics such as Gillian Beer, George Levine, and Sally Shuttleworth, as well as more recent works focusing on European writing and culture, have tended to focus the literary reception of Darwin in a single national context. Even valuable and major transnational research projects like the recent *Literary Reception of Darwin in Europe* organises its analyses according to nationality, leaving an implicit comparative analysis unelaborated.[17] This book may be read as a response to that implicit lacuna in the study of Darwin's cultural and aesthetic afterlife.

However, if this book is organised around a specific set of cultural, linguistic, and historical coordinates, its theoretical objectives are also intended to offer a set of philosophical and critical possibilities, questions, and reflections on the way in which human being (and becoming) can be represented, which is linked to, but not reducible to, national or historical origins. That is to say, while the constellation of texts — literary, theoretical, and scientific — produces and is drawn from a specific but already well-analysed set of historical and cultural coordinates, its aim is not only to begin to remap and examine them, but to draw out and explore the kinds of critical and philosophical questions that conjuncture can produce for us. The questions I am concerned with, therefore, are not only how literature has responded to Darwin's decentring of the human, but how those responses may themselves help us to re-read Darwin and through that to rethink the human.

Ineradicable Animality: Writing the Darwinian Human

If this book's generic and critical concern is with the relationship between science and literature, or between Darwinian science and the novel, the animating theoretical and philosophical impetus of this book is the question implied by Freud: what kind of human is it possible to conceive, or represent, after Darwin? If for Zola in his essay contra Balzac, naturalism, heredity, and physiology were the only proper sources for what then was the modern novel, for Freud, as I have said, the Darwinian intervention in this field of science raises questions about the ontological and epistemological status of the human as such.

Freud's reading of Darwin draws our attention to the interdependence of humanity's self-knowledge and its predication on a cosmic, biological, and psychic fantasy of exception. Yet, beyond this structural view, according to which humanity is dependent on a double-negative, its not being non-human, it also affirms that the human must be understood specifically as animal or, as Freud puts it, in terms of its 'ineradicably animal nature'. All species, Darwin writes in *On the Origin of Species* are subject to a 'constantly-recurrent Struggle for Existence' defined by a 'powerful and ever-acting means of selection', described as an economic problem:

> The struggle for existence inevitably follows from the high geometrical ratio of increase which is common to all organic beings. This high rate of increase is proved by calculation, — by the rapid increase of many animals and plants during a succession of peculiar seasons, or when naturalised in a new country.

17 *The Literary and Cultural Reception of Charles Darwin in Europe*, ed. by Thomas F. Glick and Elinor Shaffer (Continuum, 2008–14); Eve-Marie Engels, *The Reception of Charles Darwin in Europe* (Continuum, 2008).

> More individuals are born than can possibly survive. A grain in the balance will determine which individual shall live and which shall die, — which variety or species shall increase in number, and which shall decrease, or finally become extinct. [...]
>
> The slightest advantage in one being, at any age or during any season, over those with which it comes into competition, or better adaptation in however slight a degree to the surrounding physical conditions, will turn the balance.[18]

Subtly, by referring to the commonality of all organic life, this passage acknowledges humanity's connection to the animal. Yet, it also describes something constitutive about the violence, precariousness, and calculability of those universal conditions of life. We all, human and animal, share in relations of continual agonism and conflict in relation to existence, a parlous state of nature that has more in common with Hobbesian *bellum omnes contra omnium* than the notion of a divine or superior human form of life. The question of human evolution is not explicitly broached by Darwin until his later work *The Descent of Man*, published almost fifteen years later, and yet, in a way that became clear to Freud, the earlier work's implicit consequences for a conception of the human were immediately apparent to those that opposed it.[19] Most famously, in the now mythical Huxley-Wilberforce debate of 1860, Bishop Samuel Wilberforce's riposte to the Darwinian thesis in this debate, it is claimed, included a rhetorical question, aimed with mock disgust at Thomas Huxley, Darwin's representative at the meeting: '"If I may be allowed to inquire," said the Bishop of Oxford, "would you rather have had an ape for your grandfather or grandmother?"'.[20] For Wilberforce, animal ancestry was not merely distasteful, but comically unthinkable, his attempt at a humorous dismissal stemming from what today we might recognise as a deeper anxiety in relation to the threat it posed to religious authority.[21]

If in *On the Origin of Species*, the issue of human evolution is only implied, in *The Descent of Man* (1871), it is stated clearly. In the first volume of this book, Darwin provides a series of detailed comparisons between the physiological structures and

18 Charles Darwin, *On the Origin of Species by Means of Natural Selection, or the Preservation of Favoured Races in the Struggle for Life*, 2nd edn (John Murray, 1860), pp. 467–68 (hereafter referenced as *Origin*) <https://darwin-online.org.uk/content/frameset?itemID=F376&viewtype=text&pageseq=1> [accessed 15 May 2025]. This book follows the second edition of Darwin's *On the Origin of Species*, which is identical to the first edition published in 1859 but corrects misprints. Although five further editions were issued between 1859 and 1872, these include significant revisions of the text of the first edition, as Darwin sought to answer the responses of his contemporaries. For a detailed discussion of the textual history of the *On the Origin of Species*, see Richard Broke Freeman, *The Works of Charles Darwin: An Annotated Bibliographical Handlist*, 2nd edn (Archon, 1977).

19 Opposition to Darwin both before and after the publication of *The Descent of Man* often took the form of partial acceptance, with the exception of its implications for human evolution. See Martin Hewitt, 'In the Wake of the Origin', in *The Reception of Darwinian Evolution in Britain, 1859–1909: Darwinism's Generations*, ed. by Martin Hewitt (Oxford University Press, 2024), pp. 88–163.

20 J. R. Lucas, 'Wilberforce and Huxley: A Legendary Encounter', *The Historical Journal*, 22.2 (1979), pp. 313–30.

21 Richard England, 'Censoring Huxley and Wilberforce: A New Source for the Meeting that the Athenaeum "Wisely Softened Down"', *Notes and Records of the Royal Society of London*, 71.4 (2017), pp. 371–84 (p. 371), doi:10.1098/rsnr.2016.0058.

psychological capacity of 'man' and the 'lower animals', the evidence of which, it is argued, should disabuse us of the 'natural prejudice' and 'arrogance' that led our 'forefathers declare that they were descended from demi-gods'.[22] Again echoing Freud's argument about unconscious resistance, and apostrophising Wilberforce and his religious allies, Darwin in the second volume of the book diagnoses the psychic mechanism underlying religious opposition to this work: 'The birth both of the species and of the individual,' he writes, 'are equally parts of that grand sequence of events, which our minds refuse to accept as the result of blind chance'. The human ego seeks solace in religion, Darwin suggests, because of a fundamental desire to cast itself as having been 'ordained for some special purpose' and thus the product of a providential kind of teleology (*Descent*: II, 395–96). Freud's reading of Darwin implies a critique of the imaginative topology of humanity's simultaneous centrality and transcendence; Darwin offers an analysis of its narcissistic *fons et origo*.

We can read Freud's thesis on our 'ineradicable animality' and Darwin's proto-Freudian speculations on the origin of religious belief in the context of the former's later work, *Das Unbehagen in der Kultur* [*Civilisation and Its Discontents*] (1930). 'Civilisation' (*Kultur*), Freud affirms, 'describes the whole sum of the achievements and the regulations which distinguish our lives from those of our animal ancestors'.[23] Yet, for all that civilisation can 'protect men against nature', or what one might also simply call 'savagery', such an achievement is only ever a partial one in Freud's view. If civilisation can be thought as the realm of rationality and regulation, the instinctual life of the human is the 'domain *par excellence* of man's putative animality'.[24] In other words, we are never free from our (evolutionary) drives: civilisation is haunted by its unconscious and thus always undermined by the inescapable fact of our ancestral and instinctual animality. Here, Freud highlights that it is not merely the fact of humanity's connection with the animal that we resist, but the implication that the quality or condition of human life is also irradicably barbarous and unregulated. In Freudian terms, this animal haunting resides in the unconscious, the drives (*Triebe*), and instinct (*Instinkt*), as against our conscious efforts at amelioration. Yet, for Darwin, as Gillian Beer points out, no such distinction was possible (or necessary): by contrast, 'Darwin's "unconscious" signifies a permanent organic state'.[25] As such, reading Darwin's displacement of the human through a conception of an immanent state of animality and struggle is more radical even than that of Freud, for whom in *Civilisation and Its Discontents* there remains a tension between unconscious desire and attempts at regulation. In the Darwinian mode, animality does not reside in a separate, subterranean ante-chamber of the psyche; the animal and the human prowl precisely the same psychic territory, inseparable but uneasy cohabitants.

22 Charles Darwin, *The Descent of Man, and Selection in Relation to Sex*, 1st edn, 2 vols (John Murray, 1871), I, p. 32 (hereafter referenced as *Descent*) <https://darwin-online.org.uk/content/frameset?itemID=F937.1&viewtype=text&pageseq=1> [accessed 15 May 2025].

23 Sigmund Freud, *Civilisation and Its Discontents*, in *The Standard Edition of the Complete Psychological Works of Sigmund Freud*, ed. & trans. by Strachey and others, XXI (1961), pp. 57–146 (p. 88).

24 Nicholas Ray, 'Interrogating the Human/Animal Relation in Freud's Civilisation and Its Discontents', *Humanimalia*, 6.1 (2014), pp. 10–40 (p. 13), doi:10.52537/humanimalia.9926.

25 Beer, *Darwin's Plots*, p. 82.

Together then, Darwin and Freud seem to point to an aporia: the animal is the condition of (im)possibility of civilisation. It is both in relation to the animal that civilisation defines itself and from which it derives its inevitable failure. This book's three chapters can be understood as an exploration of the way in which one historically and culturally specific form of literary art, the British and French novel (in generically realist and utopian manifestations) written in response to Darwin, has engaged with and can help us explore this contradiction. The first and second chapters of this book, focusing on realist writing, explore the idea that works by Zola and Hardy instantiate a necessary confrontation with what Freud calls 'the animal' or what we might also call 'the Darwinian real'. As I will go on to demonstrate, Zola's representation of life in Second Empire France stages a drama in which human life is defined by its animality, understood as a response to a conception of Darwinian natural selection encompassing hereditary determinism and socio-economic competition. Similarly, Hardy's fiction posits a reality in which human agency is overwhelmed by impersonal forces of meaningless determination, not only the survival narratives of natural selection but also a set of rigid and oppressive biological laws of sexual relation. However, even as they construct and confront certain kinds of oppressive social-biological realities, the narratives of these novels suggest the possibility that such conditions, or realities, might be transcended, by forms of 'civilised' human activity like politics, art, and eroticism. The third chapter also addresses this tension by reading a genealogy of consciously Darwinian Utopian novels, whose function, by contrast with realist art, is thought to be propositional or speculative rather than merely representational. These works, written by a series of authors who span a historical period from Darwin's time until today, each fabulate and dramatise topoi of civilisation that are intended to transcend humanity's animality. And yet, these Utopian literary projects are marked by the impossibility of their own vision, by the ultimate collapse of the Utopian *polis*, and so like those of Zola and Hardy, might be understood as anti-utopian (or politically realist) statements on the possibility of biological transcendence.

One way of approaching this conflict between a hope of civilisational transcendence and the insistence of the Darwinian condition is found in Jacques Derrida's work on the animal. In *L'Animal que donc je suis* [*The Animal that Therefore I Am*] (2006), Derrida, alludes to Freud's three Copernican moments, arguing that it is the Darwinian attack on the formerly abyssal distinction between human and animal that elicits the most 'wounded reaction'. This is not only because of Darwinism's intimation of savagery but also because it reproduces a 'phallogocentrism' implicit in the distinction between 'symbolic' and 'imaginary'.[26] Derrida's interlocutor here is Jacques Lacan, who argues that access to the symbolic (and, therefore, human) order is defined not by a capacity to pretend, but by pretending to pretend, 'to make tracks whose deceptiveness lies in getting them to be taken as false, when in fact they are true'.[27] This double feint, Lacan suggests, amounts to being able to efface

26 Jacques Derrida, *The Animal that Therefore I Am*, trans. by David Wills (Fordham University Press, 2008), p. 136.

27 Jacques Lacan, 'The Subversion of the Subject and the Dialectic of Desire in the Freudian Unconscious', in *Écrits: The First Complete Edition in English*, trans. by Bruce Fink (Norton, 2006),

one's tracks entirely — to conceal one's origin, from oneself and others — and does distinguish humanity from the animal. For Derrida, even the pretence of pretence leaves a trace and, moreover, to ascribe to the human a capacity to 'efface its own trace', to successfully conceal its own animal ancestry, or to judge as to the success of the effacement, is symptomatic of the same myth of epistemic anthropocentrism attacked by Freud's Darwin. For Derrida, no final distinction can be made between human and animal, the latter of which is 'an appellation that men have instituted, a name they have given themselves the right and the authority to give to the living other'.[28] The human-animal distinction must remain radically indeterminate and the success of the effacement of our ancestral trace, Derrida argues, is a judgement or capacity that necessarily lies outside the purview of any being but God.

Derrida's analysis points to another direction for understanding the post-Darwinian animal-human. Neither the animal nor the human possesses the power to efface its own ancestral trace, or to judge as to its success, but humanity does, insistently, desire to do so, and therefore does possess the capacity to try and, necessarily, to fail. Indeed, if for Derrida the use of the term *animal* is a gesture of power that produces and reproduces the very notion of the animal-human divide, this same point is made by Darwin in the *Descent of Man*, where he observes that '[i]f man had not been his own classifier, he would never have thought of founding a separate order for his own reception' (*Descent*: I, 191). If by acting as its own classifier, *human* is nothing other than a signifier for humanity's ability to signify itself, what is distinctively human is the failure of this repeated effort at a final signification, at the impossibility of autobiographical self-creation. It is in the simultaneous desire for and inability to establish something transcendent via a constative called the 'human', and instead to produce a 'merely' performative entity, which creates a human that is both distinct and not from its animal others.

What is at stake in all these efforts to distinguish, categorise, efface traces, and create territories? Specifically, what contents or traits are specific to the animal which humanity seeks to repudiate or repress? Darwin speculates at the conclusion of *The Descent of Man* that the emergence of the 'half-art and half-instinct' of language not only prompts humanity to name itself, performatively to differentiate itself from its ancestors, but also to develop the faculty of morality (*Descent*: II, 390). Here, Darwin seems close to the Aristotelian and Lacanian notions of humanity as a political and symbolic animal respectively, the capacity of which to communicate and, therefore, to enter into the symbolic order, lays the ground for the development of moral concepts, the state, and 'law'. Whereas Freud sees civilisation as retaining a trace of that which it seeks to repudiate, Darwin notes that the presence of human morality represents one of the most robust arguments against his thesis of the descent of man, because it is these traits which constitute 'the greatest of all distinctions between [humanity] and the lower animals' (*Descent*: II, 392).

Yet, for Darwin, by the time he is writing *The Descent of Man*, natural selection is only one of a number of forces acting on both human and non-human life.

pp. 671–702 (p. 683).

28 Derrida, *The Animal that Therefore I Am*, p. 23.

In this work, he introduces the concept of sexual selection, according to which sexual attraction, desire, and questions of aesthetic preference are brought to bear on the propagation of species. This concept will be analysed at greater length in the following sections of this introductory chapter, and in Chapter 2 in relation to eroticism and sexual difference in Hardy's fiction. What is important here is that Darwin's work admits the possibility of determining factors on evolution other than brutal competition for survival: '[T]he moral qualities are advanced [...] much more through the effects of habit, the reasoning powers, instruction, religion, &c., than through natural selection' (*Descent*: II, 404). For Darwin, while 'social instincts' in animals and humans might emerge as a necessary evolutionary strategy, it is from this emergence that human morality is consolidated. In a sense, answering his own question (elaborated by Freud) on the ontological and epistemological status of the human, Darwin seems to return to the very anthropocentrism that his work challenges. The human, by knowing what is morally bad, distinguishes itself from the animal, which has no possible means of moral reasoning.

While Darwin in this later book seems to offer a materialist theory of moral humanism that echoes the liberal politics of his time, his earlier work on natural selection, *On the Origin of Species* provides a potentially more radical and disturbingly amoral vision of the natural world.[29] Natural selection is explicitly framed in *On the Origin of Species* as an eliminationist theory, the 'doctrine of Malthus' applied to the natural world:

> As many more individuals of each species are born than can possibly survive; and as, consequently, there is a frequently recurring struggle for existence, it follows that any being, if it vary however slightly in any manner profitable to itself, under the complex and sometimes varying conditions of life, will have a better chance of surviving, and thus be naturally selected. From the strong principle of inheritance, any selected variety will tend to propagate its new and modified form. (*Origin*: 5)

Aside from its political implications, as a theory of the relations between forms of life this Malthusian conception of nature offers no obvious sense of moral purpose or conception of good. As the biologist and philosopher Stanley Shostak puts it, '[n]atural selection is ordinarily supposed to power evolution by sifting variations in life's forms through an environment filter. The filtrate evolves; the sediment expires'.[30] The violence that is implicit in this theory is made explicit in the conclusion of *On the Origin of Species* as Darwin provides his peroration on the wider implications of his theory: 'from the war of nature, from famine and death, the most exalted object which we are capable of conceiving, namely, the production of the higher animals, directly follows' (*Origin*: 490). In natural selection, a brutal dialectic is constructed: life is borne of death; from nature's war comes peace; from famine, plenitude.

29 David Stack argues that while Darwin's work appears to biologise J. S. Mill's 'moral sense', the 'hard' hereditary determinism of such a gesture is a retroactive symptom of later eugenic and genetic thought: David Stack, 'Charles Darwin's Liberalism in "Natural Selection as Affecting Civilised Nations"', *History of Political Thought*, 33.3 (2012), pp. 525–54.

30 Stanley Shostak, *Evolution of Sameness and Difference* (CRC Press, 1999), p. 231.

Perhaps despite itself, *On the Origin of Species* hints at the political valences of this form of creative destruction. Commenting on the interdependency of species in the metaphorical and biological 'entangled bank' of life, Darwin remarks that it is widely known that 'when an American forest is cut down, a very different vegetation springs up'. Imagining himself standing amid the 'ancient Indian ruins of the Southern United States', Darwin marvels at the new life that has sprung up in their place:

> What a struggle between the several kinds of trees must here have gone on during long centuries, each annually scattering its seeds by the thousand; what war between insect and insect — between insects, snails, and other animals with birds and beasts of prey — all striving to increase, and all feeding on each other or on the trees or their seeds and seedlings, or on the other plants which first clothed the ground and thus checked the growth of the trees! (*Origin*: 75)

Unspoken in this paragraph is the occupation of Indigenous American land contemporaneous with Darwin's lifetime, a process that had catastrophic ecological and social implications.[31] The ancient 'Indian' ruins of Darwin's imagination are the metaphorical ruins of American indigenous human and non-human life.[32] The violence of insects feeding upon insects, of vegetation colonising notionally empty ground, is unwittingly linked to the violence of colonial genocide. The collapse of one civilisation makes way for new life — the abundance of life is guaranteed by constant antagonism and death, a Malthusian check on the growth of one society to make way for another. Here, as it is in Freud, the dynamic is not between *Kultur* and its potential collapse into savagery. Rather, we might recall Walter Benjamin's comment in his 'Über den Begriff der Geschichte' ['Theses on the Philosophy of History'] (1942) that '[t]here is no document of civilization which is not at the same time a document of barbarism'. Here, in Darwin's image of natural selection, barbarism and civilisation coincide.[33]

The human after Darwin, therefore, is an elusive object. It is neither animal nor god, neither moral nor amoral, and in the context of a natural order defined by universal animality, the human can only mark itself out as distinct and enduring

31 For example, the occupation of American land in California by European settlers and their fauna accelerated significantly in the 1850s as a result of the Gold Rush, demand for cattle, and overgrazing, the resultant droughts clearing the ground for 'introduced' plants to displace those indigenous to America: Alfred W. Crosby, *Ecological Imperialism: The Biological Expansion of Europe, 900–1900*, 2nd edn (Cambridge University Press, 2004), p. 154.

32 'By 1845 the Indian population of California was down to no more than a quarter of what it had been when the Franciscan missions were established in 1769. That is, it had declined by at least 75 percent during seventy-five years of Spanish rule. In the course of just the next twenty-five years, under American rule, it would fall by another 80 percent. The gold rush brought to California a flood of American miners and ranchers who seemed to delight in killing Indians, miners and ranchers who rose to political power and prominence — and from those platforms not only legalized the enslavement of California Indians, but, as in Colorado and elsewhere, launched public campaigns of genocide with the explicitly stated goal of all-out Indian extermination': David E Stannard, *American Holocaust: The Conquest of the New World* (Oxford University Press, 1992), p. 142.

33 Walter Benjamin, 'Theses on the Philosophy of History', in *Illuminations*, trans. by Harry Zohn (Houghton Mifflin Harcourt, 1968), pp. 253–64 (p. 256).

by an act of performative and ultimately errant self-creation. If there is a tension latent in the concept of 'ineradicable animality' between the necessity of civilisation and its failure, between the attempt at humanity to inscribe itself into being and its impossibility, there is also a tension within the body of Darwin's work between competing visions of human moral possibility, which index concrete social and political problems and struggles. The difference between the characterisations of human life articulated in these short readings of *On the Origin of Species* and *The Descent of Man* offer us radically dichotomous conceptions of what might be called 'human nature'. Darwin's work can offer us a biological and evolutionary basis for a renewed conception of human as the moral or law-abiding animal, which is consonant with the Aristotelian, liberal humanist conception of humanity as the reasoning animal. A troubling savagery, a sheer inescapability, of natural selection seems to stalk this optimism, in a way that might reveal human civilisation — in the form of Malthusian economic rationality or imperialist expansion, to take two examples — as itself enmeshed in a universal struggle encompassing all life, human and non-human.

To develop these questions, then, this book understands both *humanity* as a species and the *human* as a subject as entities created by language, a move that in part can be understood as having emerged from the Darwinian assault on the security of human self-authorisation. That is not to make the claim that there is no such thing as an embodied or material existence experienced collectively or individually by a kind of life we call 'human'. Rather, I work on the understanding, via Darwin, Freud, and Derrida, that in precisely the (failed) act of naming this kind of life, of calling it 'human', we construct a new being, whose unity and clarity is a function of linguistic imagination. Indeed, as Cary Wolfe's 'posthumanist' reading of Derrida argues, placing this gesture of impossibility at the core of what we understand as human, and insisting on the unstable-because-linguistic nature of our being, is the condition of possibility of 'our' lives:

> '[W]e' are always radically other, already in — or ahuman in our very being — not just in the evolutionary, biological, and zoological fact of our physical vulnerability and mortality, our mammalian existence but also in our subjection to and constitution in the materiality and technicity of a language that is always on the scene before we are, as a precondition of our subjectivity. And this means that 'what *he* calls "man,"' what 'we' call 'we,' always covers over a more radical 'not being able' that makes our very conceptual life possible.[34]

What is more, this subjection to language, in problematising the human's relation to itself and thus to the animal, opens a range of possible relations with the 'other' — including not only the non-human or the animal but also with the non-organic and the technological.[35]

This book's objective, therefore, is to examine how through the medium of one specific symbolic and historically specific form, the so-called 'Darwinian novel' — whose historical and generic coordinates I discuss further in the next section —

34 Cary Wolfe, *What Is Posthumanism?* (University of Minnesota Press, 2009), p. 88.
35 Ibid., p. 90.

responds to the Darwinian assault on humanity and, in doing so, offers through its own formal, narrative, symbolic languages the possibility of a new human articulation. For example, if for the fictional Sandoz and his authorial progenitor, Zola, biological science offers the possibility of a panoptic biological and historical realism, we might read their respective fantasies and literary efforts as attempts, or reflective of a desire, to reinscribe the human subject as transcendent narrator/observer, in response to the Darwinian revolution's assault on the fantasy of human biological sovereignty. That is to say, if Sandoz's project for scientific literary realism can be understood as an attempt to reinscribe human transcendence in an ostensible embrace of evolutionary theory, the very novel in which this idea appears can itself be read in the same manner. As such, we may trace a similar desire and effort in the works of Hardy and the three Utopian novels analysed in the following chapters, each of which respond to the Darwinian attack on humanity with different articulations or inscriptions of human life. As we shall see, these attempts are linked, not only to a desire for an abstract human removal or transcendence, but also to specific political, aesthetic, erotic, and technological aims, whose necessary failure is a challenge to our thinking of post-Darwinian human possibility.

Reading the Darwinian Novel

Three distinctive approaches have characterised literary critical engagement with the writing and ideas of Darwin: an emphasis on the influence of Darwin's work outside non-scientific and cultural spheres of activity (including literature); an illumination of Darwin's work as itself a form of literary discourse; and an exploration of the mutually interpenetrative dynamic between scientific and literary writing and thought.[36] This book, as we shall see, draws on each of these approaches, but is indebted primarily to the 'third wave' of Darwin-literature studies, spearheaded by the work of Gillian Beer, whose *Darwin's Plot's* enumerated and examined the 'two-way traffic' between 'not only [the] *ideas* but [the] metaphors, myths, and narrative patterns' of science and literature.[37] This idea enables us to see science and literature, not as the separate 'two cultures' of C. P. Snow's account (of which non-scientific knowledge is implicitly inferior), but both as part of culture as such.[38] In this study, therefore, we might see the *human* less as a problem examined in different ways by biological science and literature but as one shared by Darwin (and by Freud) and by literary writing. Such traffic as there is between these discourses indexes not a set of discrete knowledges but illuminates the internal relations of a cultural field, centred on British and French intellectual and literary discourse, in which the human has become an unstable conceptual and discursive entity.

Why, then, when the barriers between scientific, literary, and other cultural forms can be broken down in this way, organise this book around the notionally uni-

36 Devin Griffiths, 'Darwin and Literature', in *The Cambridge Companion to Literature and Science*, ed. by Steven Meyer (Cambridge University Press, 2018), pp. 62–80.

37 Ibid., p. 66.

38 C. P. Snow, *The Two Cultures* (Cambridge University Press, 2012).

directional principle of reception? Why insist on a term, or coinage, 'the Darwinian novel', which suggests a certain trajectory of discursive communication or historical relation, according to which the novel becomes Darwinian? As my elaboration of Freud's reading of Darwin suggests, I shall be working on the basis that the scientific work of Darwin's scientific writing did transform the cultural field, in which the novel is one component, in a historically and conceptually novel way. Darwin's was not the first evolutionary biological hypothesis of the long nineteenth century. It arrived in a field of thought already populated by Robert Chambers's theory of cosmological evolution in *Vestiges of the Natural History of Creation* (1844), Jean-Baptiste Lamarck's theory of acquired characteristics in *Philosophie zoologique* (1809), and his grandfather Erasmus Darwin's proto-evolutionary writing of the late eighteenth century.[39] Yet, none of these proto-evolutionary works can match either the historical and discursive consequentiality of Darwin's work or the philosophical innovation in its elaboration of a materialist and non-teleological conception of evolution, in which struggle without end was the cradle of life and not its unfortunate accident.[40] In the 1860s, numerous other major scientific publications popularised, responded to, and developed Darwin's theory of natural selection and addressed, specifically, its implications for conceptualising the human species, even if Darwin himself had yet to articulate an explicit theory of human evolution.[41] Moreover, Darwin's work also found an audience beyond the educated or scientific elite, in part because, as Raymond Williams argues, Darwin's agonistic theory of natural selection corresponded to the British public's own 'daily experience of life'.[42] As such, we may see Darwin's publication of *On the Origin of Species* as effectively splitting the nineteenth century in two, creating a post-Darwinian epoch from which point it becomes impossible to write as if one belonged to a prelapsarian ante-Darwinian innocence, even if — as many in the nineteenth century sought to do — one's purpose was to reject, either wholly or in part, Darwin's conception of evolution and its implications for humanity.

For precisely this reason, the phrase 'the Darwinian novel' seeks to specify a subset of post-Darwinian literary texts that do, with varying levels of consciousness and explicitness, engage with Darwin's work, with varying degrees of accuracy and fidelity to the original ideas and writing. This notion, as we shall see, can be supported by evidence of reception or conscious engagement in these novelists' literary and non-literary writing. The 'Darwinian novel', then, rather than seeking

39 Robert Chambers, *Vestiges of the Natural History of Creation* (John Churchill, 1844); Jean Baptiste Lamarck, *Zoological Philosophy: An Exposition with Regard to the Natural History of Animals*, trans. by Hugh Elliot Macmillan (University of Chicago Press, 1984).

40 Janet Radcliffe Richards, *Human Nature after Darwin: A Philosophical Introduction* (Routledge, 2005), p. 17; Daniel Dennett, *Darwin's Dangerous Idea: Evolution and the Meanings of Life* (Penguin, 1996), p. 73.

41 'A string of publications addressed, in the title of Huxley's 1863 intervention, *Man's Place in Nature*. John Lubbock's *Prehistoric Times* (1865), and in particular Charles Lyell's *Antiquity of Man* (1863), publicised the archaeological discoveries of the previous decade and their demonstration of the great age of humanity': Hewitt, 'In the Wake of the Origin', pp. 88–90.

42 Raymond Williams, 'Social Environment and Theatrical Environment', in *Culture and Materialism: Selected Essays* (Verso, 2005), pp. 88–104 (p. 92).

to establish a generic category, seeks instead to register the difference between all post-Darwinian literature (or novels) where we might find Darwinian themes or narrative substrates, and those novels of the post-*Origin of Species* period where some explicit evidence of reception is present. Moreover, as we shall see, the Darwinian novel is not a homogenous formal, generic, or narrative entity, nor does it offer a straightforward image of reception and literary influence: the literary materiality of these texts often problematises or contradicts the evidence of historical reception, suggesting either an unconscious or unacknowledged form of engagement with different Darwinian ideas which escape or at least problematise historical constructions of authorial intention.

This tension in the Darwinian novel between the para-literary evidence of reception and the literary material is emblematised by the material presented in the first chapter, which offers one account of the complex reception and translation history of Darwin in Zola's work in the context of nineteenth century France and, in turn, explores the specifically political interpretation of Darwinism present in Zola's fiction which both reflects and complicates that genetic story. In the second chapter, the case of Hardy's reception of Darwin seems empirically simpler than that of Zola, since it lacks the mediatory complexities of translation and cultural transfer. Yet, as I will show, *A Pair of Blue Eyes* and *The Return of the Native* complicate a certain critical orthodoxy — according to which Hardy's reception of Darwin is defined wholly by a concern with struggle and existential emptiness — by responding instead to Darwin's lesser-known theory of sexual selection and its narratives of courtship and desire. The final chapter explores another complex image of reception in three Utopian novels by three different authors, Samuel Butler, Aldous Huxley, and Michel Houellebecq. This traces a longer genealogy of reception from Darwin's time to our own and demonstrates the way in which the reception of Darwin by a contemporary author such as Houellebecq takes place via the mediation of Darwinian thought and narratives in works such as those of Aldous Huxley, who, in turn, was influenced by Butler. Such intertextual connectivity echoes suggestively the genealogical narrative of evolution as such and gesture towards a futurity with which the Utopian novel is often constitutively concerned. Yet, in these works, Darwinian thought acts as both the condition of possibility of utopian futurity and a connectivity between past, present, and future, as well as the breakdown of those possibilities and connections.

How does this approach — concerned at once with history, language, and literature's theorisation of the human — differ, if at all, from previous ways of engaging with literature and its relation to Darwinian science? How do others, concerned with literature, read Darwin? And how do others read those who read Darwin? One method is to apply the methods and principles of Darwinian evolution to the study of literature. In his 1907 essay 'On the Application of Evolutionary Principles to Art and Literature', John Addington Symonds argued that Darwin's theory of evolution afforded the basis for a natural history of the rise and fall of literary and cultural types.[43] In France, Ferdinand Brunetière articulated

43 John Addington Symonds, 'On the Application of Evolutionary Principles to Art and Literature',

a markedly similar position in his book-length study *L'Évolution des genres dans l'histoire de la littérature* [*The Evolution of Literary Forms*] (1890). In it he alluded to the advances made by Darwin in the field of biological classification, and used this as the basis to trace the growth, decay, and extinction of a range of literary genres in European history.[44] Today a similar approach to Darwinian literary criticism is taken by self-described 'Literary Darwinists', whose work seeks to bring to bear a synthesis of Darwinian natural selection and contemporary genetics on the study of literature.[45] Its fundamental epistemological thesis is the claim that 'all knowledge about human behaviour, including the products of the human imagination, can and should be subsumed within the evolutionary perspective'.[46] The historical echoes between these two approaches, it has been argued, springs from a contemporary positivistic relation to scientific knowledge which renews Victorian anxieties about epistemological trustworthiness. If for Darwin, religious belief played the role of defending the human ego against its own dissolution, literary Darwinian fantasies of a unified conception of science, both in the nineteenth and twenty-first centuries, play the same role.[47] We may hear too the echo of Sandoz's claim in *L'Œuvre* that science is the only possible basis for a new universal and universalisable scientific knowledge of humanity.

This book, it should be clear, does not aim to attempt such a scientific or pseudo-scientific approach to the study of literary creation and discourse. Yet, such approaches do attest to an important facet of the cultural reception of Darwin. For although the studies by both Brunetière and Symonds reference Darwin explicitly, their respective critical approaches to literary history owe as much to teleological pre-Darwinian accounts of evolution, such as the work of the then widely-read evolutionary social theorist Herbert Spencer and the French naturalist Jean Baptiste Lamarck. It might also be pointed out that their narratives of cultural growth, decay, and extinction can be located in a longer philological perspective and wider 'evolutionary' tendency in history derived from Hegel. Yet, while they may not be accurately or honestly Darwinian, by signalling their work as 'Darwinian', these early efforts to account for and make use of Darwin index the wider philosophical and cultural significance (and prestige) of Darwinian evolution, in Britain and in

in *Essays: Speculative and Suggestive*, 3rd edn (Smith, Elder, 1907), pp. 27–57.

44 Ferdinand Brunetière, *L'Évolution des genres dans l'histoire de la littérature: leçons professées à l'École normale supérieure* (Hachette, 1890).

45 'Literary Darwinism' is the title given to the discipline by its de facto leader, Joseph Carroll, whose works have been foundational in the emergence of the critical field. See Joseph Carroll, *Literary Darwinism: Evolution, Human Nature, and Literature* (Routledge, 2004); and *Reading Human Nature: Literary Darwinism in Theory and Practice* (State University of New York Press, 2011). Brian Boyd is an example of a critic whose outlook is less strident than that of Carroll and who describes himself more modestly as an 'evocritic', but his work is still guided by the same commitment to advancing human knowledge through 'Darwinian' analyses of literary works. See Brian Boyd, *On the Origin of Stories: Evolution, Cognition, and Fiction* (Belknap Press of Harvard University Press, 2010).

46 David DiSalvo, 'What Is Literary Darwinism? An Interview with Joseph Carroll', *Neuronarrative*, 27 February 2009 <https://neuronarrative.wordpress.com/2009/02/27/what-is-literary-darwinism-an-interview-with-joseph-carroll/> [accessed 1 May 2016].

47 Vanessa L. Ryan, 'Living in Duplicate: Victorian Science and Literature Today', *Critical Inquiry*, 38.2 (2012), pp. 411–17, doi:10.1086/662752.

France, by the late nineteenth and early twentieth centuries — an exploration of which cultural field is this book's objective.[48] Darwin's name, as I have argued elsewhere, is used frequently, and in different ways, to provide various kinds of cultural and scientific cachet, for which a faithful or rigorous relation to Darwin's work is not necessary.[49] These examples of literary Darwinian approaches may also testify to the way resistance to Darwin's assault on anthropocentrism can also take the form of adoption. Darwinian evolution — whether as a theoretical framework or signifier — is used in these examples to reconstitute a space of epistemological dominance for the human subject.

Yet another contradiction implicit in these approaches is that, in order to construct knowledge about literature within a universal and undifferentiated field, they rely on making a strict distinction between what is scientific and therefore testable knowledge (various forms of Darwinian science) and unscientific and unverifiable kinds of truth (in literature or art). This differentiation between scientific and literary discourse or knowledge is implicitly echoed in some twentieth-century criticism (what has been called the 'second wave' of Darwin studies), such as that of Lionel Stevenson's *Darwin among the Poets,* in which Darwin is constructed as one exceptional case of a specifically scientific knowledge penetrating the 'ivory tower of poetry'.[50] In a similar vein, William Irvine writes in 'The Influence of Darwin on Literature' (1959) that Darwin's work 'in the strict literary sense' should not be considered prosodic or poetic but a strictly scientific kind of writing.[51] While neither Irvine nor Stevens make an explicit case for universal epistemological 'consilience' in the way that their predecessors and descendants in Darwinian literary criticism do, they insist instead on a discursive separation between scientific writing and literary writing.

Rather than search for a unified knowledge about literature through Darwinian science, or to separate in a final way the discourses of Darwinian science and literature, this book instead is concerned with elaborating the kind of knowledge that literature can produce through its discursive, historical, and theoretical interpenetration with the work of Darwin. This approach assumes the primacy of 'discourse' over knowledge, or rather, makes the case for the discursive nature of knowledge. It is also an explicitly historicist approach, which assumes that whatever theoretical knowledge is rendered by the critical encounter of scientific and literary works is most fruitfully understood in relation to its historical and cultural context. These epistemological and historical dispositions can be located in the pathbreaking work on Darwinian narrative that emerged in the 1980s, such as that of Beer's *Darwin's Plots*, as well as many works thereafter, which examine the 'inter-related

48 René Wellek and Stephen G. Nichols, *Concepts of Criticism* (Yale University Press, 1963), pp. 37–53.

49 Niall Sreenan, 'In the Name of the Father: Darwin, Scientific Authority and Literary Assimilation', in *From Shakespeare to Autofiction*, ed. by Martin Procházka (UCL Press, 2024), pp. 91–108.

50 Lionel Stevenson, *Darwin among the Poets* (Russell & Russell, 1963), p. 140.

51 William Irvine, 'The Influence of Darwin on Literature', *Proceedings of the American Philosophical Society*, 103.5 (1959), pp. 616–28 (p. 619).

network' of the natural sciences and literature.[52] This efflorescence of work has precipitated its own cultural poetics of Darwinism, or, as George Levine puts it, an academic 'Darwin industry'. This is a field replete with studies of Darwin's influence on the Victorian age and beyond, and includes studies not only of the canonical nineteenth-century novelists in Britain whose debt to Darwinism is now a critical commonplace (including George Eliot, Thomas Hardy, H. G. Wells, and others), but also studies of transnational reception (across Europe and beyond) as well as other cultural and literary forms beyond the novel, such as dance and theatrical form.[53]

While such a profusion of work suggests a number of formal possibilities for the study of the cultural afterlife of Darwin, this present study shares with Beer and critics such as Levine and Sally Shuttleworth a focus on narrative fiction. For Beer, evolution's 'preoccupation with time and with change' has particular resonances for the 'problems and processes of narrative'. Indeed, '[e]volutionary theory,' she writes, 'is a form of imaginative history' — and it is productive here to understand the term according to its root, *historia*, which denotes not narratives of the past but narrative accounts *tout court*.[54] Focusing on the novel form within and beyond the nineteenth century is not only a reflection of the emergence of the novel as a popular literary form in nineteenth-century Britain and Europe. It also follows the premise that Darwin's impact on writing can be found especially clearly and suggestively in the historically specific narrative forms and structures of nineteenth-century realism. As Levine has argued, the realist novel in the nineteenth century (both in Britain and France) responds to, is influenced by, and shares an objective with growing scientific aspirations to a representation of the real. If this is explicitly found in Zola's work in the French context, as I have already shown, the Victorian novel incorporates Darwinian concepts less in the form of 'ideas' than in its formal rhetoric, narrative causality, and the relationships it constructs.[55] The presence of Darwinian narrative causality is addressed in this book's second chapter on Hardy's novels, where the question of fatalism is considered in relation to a Darwinian logic of determinism. The inclusion of three Utopian novels in this study offers a counter-example to the case of realist fiction, demonstrating that questions of evolutionary form and representation are not limited to explicitly or generically realist writing. While nineteenth-century realism is concerned with the representation of reality, with narrative causality, and determinism, these Utopian novels can be linked to this project by their preoccupation with their opposites: speculation about what

52 Sally Shuttleworth, *George Eliot and Nineteenth-Century Science: The Make-Believe of a Beginning* (Cambridge University Press, 1984), p. 16.

53 George Levine, 'Foreword', in Beer, *Darwin's Plots*, pp. ix–xiv (p. ix). See, for example: George Levine, *Darwin and the Novelists: Patterns of Science in Victorian Fiction* (Harvard University Press, 1988); Martin Fichman, *Evolutionary Theory and Victorian Culture* (Humanity Books, 2002); *The Evolution of Literature: Legacies of Darwin in European Cultures*, ed. by Nicholas Saul and Simon J James (Rodopi, 2011); Virginia Richter, *Literature after Darwin: Human Beasts in Western Fiction 1859–1939* (Palgrave Macmillan, 2011); Rae Beth Gordon, *Dances with Darwin, 1875–1910: Vernacular Modernity in France* (Ashgate, 2009).

54 Beer, *Darwin's Plots*, pp. 5–7.

55 Levine, *Darwin and the Novelists*, p. 13.

might be and a break with established causal patterns of history and change.

Yet, as I have already said, my aim in this book is not only to establish whether and how these works have been influenced by Darwin, or only to further illuminate the 'web of connection' that defines the relationship between science and literature (an approach that itself owes something to the image of 'entanglement' in Darwin's writing).[56] It is also to read the relationship of reception backwards and ask: how might our own reception or interpretation of Darwin's work change in the light of its literary afterlife? How might the 'Darwinian novel' allow us to think the question of the human in new ways? To ask these questions is simply to acknowledge how the act of tracing reception in the traditional historical sense, of tracking influence chronologically from the past to the present, inevitably involves a certain kind of 'afterwardsness', or *Nachträglichkeit*, according to which the knowledge we gain at the end of that process retroactively transforms the meanings of that which preceded it.[57] For Freud, this kind of retroaction was an unconscious process: an event experienced in the past often does not become traumatic or significant for the subject until called forth by some new knowledge or momentous experience in the present. Is this not precisely what Freud is pointing to in his comments on the scientific revolutions of Copernicus, Darwin, and psychoanalysis? It is only the hostility, rejection, or disavowal that these major theoretical interventions produce that allow us to grasp, in retrospect, their fundamentally radical import.

It is one of the aims of this book, therefore, to show how the shock of Darwin's intervention in science can be understood through the retrospective critical and theoretical lens of the works of literature that came after it. In other words, by working *ex post facto*, by becoming more attuned to the ways in which literature *after* Darwin reads and responds to Darwinian thought, we develop new ways of relating to and interpreting Darwin's work and its dethroning of the human for the future. If for Darwin, 'the struggle was to explain how we can both be part of nature *and* have the developed capacities to remake that nature', the challenge in this work is to continue that struggle by reading those whose literary responses to Darwin may offer a vision of human nature remade.[58] Here, the 'one culture' thesis, in which science and literature share a discursive space allows us not only to explore the question of the human as it is explored in a culture that includes science and literature, but also as a space in which to articulate the knowledge produced by literature about the human after Darwin.

56 Ibid., p. viii.

57 Anne Golomb Hoffman, 'The Child in the Adult: Narrative and *Nachträglichkeit* in Henry James and Freud', *Psychoanalytic Quarterly*, 94.1 (2025), pp. 93–122, doi:10.1080/00332828.2024.2442117.

58 Devin Griffiths and Deanna K. Kreisel, 'After Darwin: Ecology, Posthumanism, and Aesthetics in the Twenty-First Century', in *After Darwin: Literature, Theory, and Criticism in the Twenty-First Century*, ed. by Devin Griffiths and Deanna K. Kreisel (Cambridge University Press, 2023), pp. 1–19 (p. 14); this edited volume of essays offers a suggestive instantiation of a theoretical project that takes on the challenge of theory and criticism in a post-Darwinian epoch.

Darwinism between Eros and Thanatos

The human after Darwin seems to flicker between savagery and civilisation, a being that is unavoidably subjected to and part of a state of nature, as well as defined by a partial capacity to transcend that state via the affordances of civilisation. In that light, I have postulated that literature's symbolic materiality may work to write and re-write a post-Darwinian version of humanity — without recourse to new forms of exceptionalist fantasy — and that we might find particularly historically, critically, and philosophically suggestive instantiations of that in a series of novels written in response to the Darwinian revolution in biological science (and its wider shift in our cultural gestalt). Before finally moving on to the literary critical project of this book and the visions of humanity we find in Zola, Hardy, and three Utopian fictions, this final section of the Introduction elaborates further on some of the important theoretical and historical coordinates of this divided Freudo-Darwinian human which inform and will be developed in the three chapters that follow.

If one of Darwin's major innovations in *On the Origin of Species* is to claim via the theory of natural selection that all life is driven by a specifically biological 'struggle for life', it is also true to say that warfare, famine, death, and struggle have been represented in Western thought as a natural way of being at least since Thomas Hobbes described the state of nature as a 'war of all against all' and life within it as 'nasty, brutish, and short'.[59] Like Freud, Hobbes offers a solution to natural savagery in the form of a regulatory function. Yet, Hobbes's Leviathan, in which the animality of the multitude is sublimated in the form of the state's monopoly on violence, is merely redistributive: the beast stalks away from the heart of the multitude and takes up residence in the sovereign state. A more pessimistic and so notionally realistic conception of the nature of life may be found in the work of Arthur Schopenhauer. In *Die Welt als Wille und Vorstellung* [*The World as Will and Representation*] (1818–19), Schopenhauer claimed that all life was characterised by its agonistic relation to death:

> Thus a constant struggle is carried on between life and death, the main result whereof is the resistance by which that striving, which constitutes the innermost nature of everything, is everywhere impeded. It presses and urges in vain; yet, by reason of its inner nature, it cannot cease.[60]

One of a number of scholars to draw parallels between this view and natural selection, the anthropologist David Asher, in 1871, detected the echoes of Schopenhauerian nihilism in Darwinian natural selection: 'Darwin's speculations, based on the inductive method, have now corroborated the deductive theory of Schopenhauer's'.[61] Darwin had established in scientific and biological terms what

59 Thomas Hobbes, *Leviathan*, ed. by Richard Tuck (Cambridge University Press, 1991), pp. 88–90.

60 Arthur Schopenhauer, *The World as Will and Representation*, trans. by E. F. J. Payne, 2 vols (Dover, 1966–69), I, p. 349.

61 David Asher, 'Schopenhauer and Darwinism', *Journal of Anthropology*, 1.3 (1871), pp. 312–32 (p. 313). See also, Gustav Weng, *Schopenhauer-Darwin: Pessimismus oder Optimismus? Ein Beitrag zur Fortschrittsbewegung* (Ernst Hofmann, 1911); an English review and summary of this can be

Schopenhauer's work postulated in metaphysical ones. And if for Darwin 'struggle' is not an explicitly metaphysical or ontological condition, natural selection shares with Schopenhauerian nihilism a sense of its ceaselessness — a relentless dialectic according to which the will to live is both constrained and necessitated by finitude. For Hobbes and Schopenhauer, the immanence of struggle and the suffering that accompanies it are political and metaphysical concepts. Thomas Malthus's work, which was so influential on Darwin's theory of natural selection, elaborates a theory of struggle in economic terms. Contrary to a Godwinian conception of human nature and 'the perfectibility of man', Malthus argued at the beginning of the nineteenth century that while human population continued to increase, an inability to provide resources and food at the same rate of growth would lead inevitably to a conflict over resources and the necessary failure of the weak.[62] For Malthus, while food production and population increase were elastic, 'the cruel customs of humanity' were a constant: together they act as an internal check on human progress, re-entrenching suffering and competition according to a fixed mathematical logic.[63]

So, Darwin's theory of natural selection has a series of intellectual precedents, which should prevent us from seeing the agonistic worldview of *On the Origin of Species* as arriving *ex nihilo* into a Victorian intellectual field. However, as Karl Marx observed, Darwin's vision of struggle seemed to naturalise a specifically ideological and historically contingent conception of social relations and economic production: 'It is remarkable how Darwin recognizes among beasts and plants his English society with its division of labour, competition, opening-up of new markets, "inventions", and the Malthusian "struggle for existence". It is Hobbes' *bellum omnium contra omnes*'.[64] For Raymond Williams, who observed that this vision of life was already present in everyday social relations in the Victorian period, such a view also helped to frame, in biological terms, the incessant warring of imperialist nations as an expression of the deepest aspect of nature itself.[65] War, conflict, struggle: these were not the incidental phenomena of historical circumstance, nor the result of ideological precepts, economic relations, or imperialist ambition. Nature was at war and from this, all other struggles flowed.

This account of natural selection is a theory of Darwinian evolution under the sign of Thanatos. For Freud, in *Jenseits der Lustprinzips* [*Beyond the Pleasure Principle*]

found in J. H., 'Schopenhauer-Darwin: *Pessimismus oder Optimismus*', *Nature*, 85.403 (1911), p. 403, doi:10.1038/085403a0; Arthur O. Lovejoy, 'Schopenhauer as an Evolutionist', *The Monist*, 21.2 (1911), pp. 195–222.

62 William Godwin, *An Enquiry Concerning Political Justice, and Its Influence on General Virtue and Happiness*, 2 vols (Robinson, 1793).

63 T. R. Malthus, *An Essay on the Principle of Population, or, A View of Its Past and Present Effects on Human Happiness with an Inquiry into Our Prospects Respecting the Future Removal or Mitigation of the Evils Which It Occasions*, 1st American edn (R. C. Weightman, 1809); Piers J. Hale, *Political Descent: Malthus, Mutualism, and the Politics of Evolution in Victorian England* (University of Chicago Press, 2014).

64 Letter from Karl Marx to Friedrich Engels, 18 June 1862, in *Karl Marx and Friedrich Engels: Selected Correspondence*, ed. by S. W. Ryazanskaya (Progress, 1965), p.128.

65 Williams, *Culture and Materialism*, p. 93.

(1920), 'the predicate of being "instinctual"' was 'related to the compulsion to repeat' and its destructive and intrinsically anti-progressive characteristics.[66] Organic life, Freud speculates, is constituted by repetitions: of the kinds of repeated organic phenomena, like reproduction, migration, and self-preservation associated with growth and development. However, such instincts, in this account are components of a deeper 'conservative' drive, the aim of which is precisely not to change or 'evolve' but to return to an inanimate state: to return to death. What appear as facets of a self-preservative drive are, for Freud, only symptomatic detours before this final aim, an attempt by the organism to 'die in its own fashion'. If for Freud, then, death is the aim of all life, and the destruction, masochism, and sadism that characterise human life are symptomatic of that underlying compulsion, in Marx's Hobbesian conception of Darwinian natural selection, life is the objective of the organism or species, which it achieves at the expense of its rivals.

Yet, for some theorists of evolution, progress and eliminationism are not mutually exclusive. For theorists such as Francis Galton, the founder of British eugenics, and Herbert Spencer, the evolutionary social theorist and philosopher, the 'survival of the fittest' provided paradoxical support contra Malthus for a belief in the possibility of human amelioration. For Galton, in *Hereditary Genius* (1869), selection processes offered the possibility to improve the 'stock' of the human species: 'The processes of evolution are in constant and spontaneous activity, some pushing towards the bad, some towards the good'; the role of the scientifically enlightened society was to be judicious in intervening by 'checking the former and giving free play to the latter'.[67] Spencer, for his part, was strictly-speaking an exponent of a teleological form of Lamarckian evolutionism. Yet, he too argued that natural selection (or the 'struggle for existence') offered the key to social progress, in his case by offering scientific support for free-market capitalism and individualism against the rising threat of 'social revolution'. Foreshadowing the dialectic suggested in Freud's *Civilisation and Its Discontents*, Spencer argued that the modern affordances of the capitalist and industrial economy, the 'appliances and products distinguishing civilisation from savagery', owe their genesis not to social cooperation or labour but to the 'struggle for existence'.[68] Spencer is cognisant of the 'evils' incentivised by the Darwinian free market: dishonesty, adulteration, greed, corruption, deception, fraud, and a system that disproportionately rewards those who 'regulate and superintend' as against 'the actual workers'.[69] Yet, such symptoms were merely 'relative evils', offset by the social and economic progress afforded by competition.

66 Freud uses the term *Triebhaft*, that is to be compelled by *drive*, as opposed to one derived from the term *Instinkt*. Nevertheless, Strachey's translation choice succeeds here in denoting a form of unconscious psychic life and its relation to organic life. See Sigmund Freud, *Beyond the Pleasure Principle*, in *The Standard Edition of the Complete Psychological Works of Sigmund Freud*, ed. & trans. by Strachey and others, XVIII (1955), pp. 7–64 (p. 36).

67 Francis Galton, *Hereditary Genius: An Inquiry into Its Laws and Consequences*, 2nd edn (Macmillan, 1892) p. xxvii.

68 Herbert Spencer, 'From Liberty to Bondage (1891)', in *The Man versus the State: With Six Essays on Government, Society, and Freedom* (Liberty Classics, 1981), pp. 487–519 (p. 491).

69 Ibid., p. 514.

While 'the war of all against all' seems distinct in its naturalist pessimism from the whiggish liberalism of Galton and Spencer, these thinkers share a conviction that human animality can be transcended. This is not to be achieved, however, by a flight from savagery, violence, or suffering, but their embrace. To avoid 'continual fear, and danger of violent death', Hobbes's theory prescribed the necessity of separating humanity from the cruelty of the state of nature through the invention of a sovereign power: an autonomous enforcer of supra-natural civil law to which humanity ceded some of its more animal freedoms.[70] Spencer and Galton posit analogous, but innovatively biopolitical, modes of enforcement and control. Galton's theories of eugenics sought to shape evolutionary development with reference to a normative, ideal vision of the human.[71] For Spencer, the enforcement of free-market economics and competitive social relations ensured only the best individuals would survive the melee to create an ideal society.[72] Each theorist proposed a type of self-separation, an incision in the very fabric of human evolution, achieved, paradoxically, by an appeal to the law of competitive evolution, the conditions of which are what the human subject and species should aim to transcend. To escape the war of nature, to enact the definitive break between the civilised and the animal, the human must first embrace its animality.

Here, Freud's notion of the death drive is apt. Progress, the creation of a new human subject, species, or society is achieved through a repetition — repeated violence, repeated struggle, repeated elimination of the weak. Yet, such a nihilistic embrace of the Darwinian maxim of the struggle for life, paradoxically, does not offer a vision of the human as animal, or indeed contribute to the re-conceptualisation of a new kind of post-Darwinian subject and species. Instead, the human emerges paradoxically recentred both as a helpless victim of biological fate and as a rational economic or biological agent capable of guiding the direction of evolutionary change. It is in this way that Darwinian evolutionary thought can seem to be endorsed at the same time as its most radical implication — the decentring of human life — can be resisted.

There is another way of relating to the Darwinian revolution. It is hinted at in Freud's *Beyond the Pleasure Principle* and its allusion to the 'germ plasm' theory of heredity. Freud saw in his theorisation of the death instincts, 'which seek to lead what is living to death', and the sexual instincts, Eros, aimed at reproduction and renewal, an analogy in the German evolutionary biologist August Weismann's

70 Helen Thornton, *State of Nature or Eden? Thomas Hobbes and His Contemporaries on the Natural Condition of Human Beings* (University of Rochester Press, 2005), pp. 141–44.

71 'There is nothing either in the history of domestic animals or in that of evolution to make us doubt that a race of sane men may be formed who shall be as much superior mentally and morally to the modern European, as the modern European is to the lowest of the Negro races'. Thus, Galton argued, eugenic efforts should be directed at emulating this superior 'breed' of human: Galton, *Hereditary Genius*, p. x.

72 Richard Hofstadter dedicates a chapter to Spencer in his work *Social Darwinism in American Thought* (1944) and argues that his free-market vision of nature was well-suited to an already extant aggressively individualist ideological atmosphere in the American imaginary: Richard Hofstadter, *Social Darwinism in American Thought*, rev. edn (Beacon Press, 1955), pp. 31–50.

conception of heredity.[73] Weismann makes the distinction between the 'soma', which is mortal, and the germ-cells, as a potentially immortal substance that acted as the conduit through which reproduction could occur:

> The difference between the 'body' in the narrower sense (soma) and the reproductive cells was also emphasised, and it was maintained that the germ-cells alone transmit the reproductive substance or germ-plasm in uninterrupted succession from one generation to the next, while the body (soma) which bears and nourishes the germ-cells, is, in a certain sense, only an outgrowth from one of them.[74]

For Freud, Weismann's theory suggests that biological life is not wholly dominated by the death instinct, Thanatos. It is instead locked in in tension with a life drive, or Eros — the realm of those symptomatic detours taken by the subject to ensure their return to non-being happens in a way that it finds acceptable. Here, post-Darwinian human existence is not entirely thought to be determined by a morbidly mechanistic process of conflict and extinction, or by death as the necessary telos of all life. Here too, in what Freud calls the 'sexual instincts' there is a kind of immortality in the creativity of Eros; there is divergence, pleasure, and deferral in this biological, if not explicitly Darwinian, conception of the human animal.

The germ of evolutionary optimism found in Freud's work is further developed by (the famously counter-Freudian thinker), Gilles Deleuze. Like Freud, Deleuze observes that the full significance of Darwin's work has yet to have been fully appreciated. However, unlike Freud, for whom the question of the dissolution of the human generates a series of questions about the problems of animality (only partially answerable by a speculative, Weismannian account of sexual instinct), for Deleuze, the dissolution both of species and subjects is the necessary condition for a new (and potentially post-human) vital philosophy of difference. Writing in *Difference and Repetition*, Deleuze identifies that the fundamental innovation of the 'Copernican Revolution of Darwinism' was the inauguration of the 'thought of individual difference'. For Deleuze, natural selection does play 'the role of a principle of reality, showing how differences 'become connected to one another and accumulate in a given direction' — or, to put it in Darwin's terms, how 'variations', however slight, aggregate in a direction to create groups, varieties, and species.[75] Yet, natural selection's effect is not only to generate growth, that is, to produce connected aggregates or distinct forms or species. Evolution can also be understood as the relentless biological promulgation of difference as such: the 'differentiation of difference (survival of the most divergent)'. Here, difference is not understood as a characteristic or feature of an organism or species, but as that which pre-exists these forms. Darwinian natural selection not only decentres our image of ourselves as clearly differentiated human subjects or species, but as stable subjects and species as such: 'The great taxonomic units — genera, families, orders and classes — no

73 Freud, *Beyond the Pleasure Principle*, p. 46.

74 August Weismann, *The Germ-Plasm: A Theory of Heredity*, trans. by W. Newton Parker and Harriet Rönnfeldt (Scribner, 1893), p. 9.

75 Gilles Deleuze, *Difference and Repetition*, trans. by Paul Patton (Continuum, 2004), pp. 248–50.

longer provide a means of understanding difference by relating it to such apparent conditions as resemblances, identities, analogies and determined oppositions'.[76] In place of the human as an identity constructed in opposition to nature or, by analogy, to the divine, now stands ontological difference and its proliferation.

For Deleuze, Darwin's theory of natural selection releases the notion of difference from its transcendental constraints, by indicating that processes of differentiation precede differences themselves, and that consequently, difference is an immanent pre-condition for life as such: species, individual organic beings, sexuation, each 'turn around individual difference, not vice versa'. Like Freud's speculative account of organic drive in *Beyond the Pleasure Principle*, Deleuze draws on Weismann to offer an explanation of the way difference is produced via (and not merely selected for) sexed reproduction. Where others may perceive 'free-floating or unconnected' difference, that is, forms of variability where it seems selection is no longer operative, Deleuze instead finds difference's 'natural cause' in Eros: sexed reproduction and, in turn, its production of sexual differentiation.[77]

If in Freud, then, apostrophising Darwinian natural selection, the conflict between Eros and Thanatos in the organic is characterised by the relentless return of the death instinct, for Deleuze, organic life possesses a creative capacity that exceeds the horizon of extinction and the survivalist mechanics of reproduction:

> Univocity of being, in so far as it is immediately related to difference, demands that we show how individuating difference precedes generic, specific and even individual differences within being; how a prior field of individuation within being conditions at once the determination of species of forms, the determination of parts and their individual variations.[78]

Natural selection, therefore, is part of a unitary field of being in which difference is prior to its differentiated objects: only one of the mechanisms of 'determination' that create species, variations, and individuals. It is preceded by a form of difference within being, a difference that inhabits life prior to the modes of selection that operate there. As such, Darwin's theory of natural selection is significant in providing a means for this philosophical inversion of a certain biological orthodoxy. For Darwin, in a Deleuzean view, neither species, variations, nor individuals function as the agents of organic change, but are themselves emergences from a supra-human (and supra-species-level) force of 'becoming'.

The work of Elizabeth Grosz, feminist philosopher and theorist, takes up this understanding of Darwin and expands upon the question of organic sexual difference and its relation to being, as well as to political liberation. For Grosz, Darwin 'not only developed the theory of natural selection into a scientific research paradigm of unparalleled fruitfulness and success for nearly a century and a half, [but] also produced a philosophical framework whose resonances have still not been properly understood, even today'.[79] She takes up Deleuze's understanding of natural

76 Ibid., p. 248.
77 Ibid.
78 Ibid., p. 38.
79 Elizabeth Grosz, *The Nick of Time: Politics, Evolution, and the Untimely* (Duke University Press,

life, through the prism of Henri Bergson's biological vitalism and Luce Irigaray's feminist theory, and argues that Darwin brings duration into the thinking of life: transcendent categories of species are replaced with durational, time-bound entities emerging from the present and dissolving into the future. Consequently, like Deleuze, she argues that nature 'in spite of its scientific reduction to closed systems operating according to predictable laws, also carries, as it were in secret, duration, flux, becoming, at its very core'.[80] Going far beyond Freud's observation that a Darwinian conception of nature displaced humans from its own imagined centre, Grosz, along with Deleuze, sees nature as existing without a stable centre-point, telos, or ontological fixity. Such a view, for Grosz, has implications for theorising feminist emancipation beyond a politics of identity.[81] She argues that a properly revolutionary feminist project must register materiality and the biological, without capitulating to the view that Darwinian evolution imposes an exhaustive description of the natural world or that it pre-determines all possible action. Natural selection's dismantling of fixed, essential nature by itself, however, does not suffice, for it does not address the question of sexuality or reproduction. To complete her elaboration of a post-structural, feminist Darwin, however, she does not return to Weismann as Freud and Deleuze do, but instead turns towards Darwin's other major contribution to evolutionary biology: sexual selection.

Sexual selection, Darwin writes, 'depends on the advantage which certain individuals have over other individuals of the same sex and species, in exclusive relation to reproduction' (*Descent*: I, 256). Traits which have no immediate bearing on an organism's capacity to survive in the 'struggle for life' are developed in sexually dimorphic species in a competition to attract mates and to gauge attractiveness:

> There are many other structures and instincts which must have been developed through sexual selection — such as the weapons of offence and the means of defence possessed by the males for fighting with and driving away their rivals — their courage and pugnacity — their ornaments of many kinds — their organs for producing vocal or instrumental music — and their glands for emitting odours; most of these latter structures serving only to allure or excite the female. That these characters are the result of sexual and not of ordinary selection is clear, as unarmed, unornamented, or unattractive males would succeed equally well in the battle for life and in leaving a numerous progeny, if better endowed males were not present. (*Descent*: I, 257)

Of course, Darwin still has in mind here a kind of competition, which is not only androcentric (and phallic) in its conception of male competition for female sex objects but also produces various morphologies of assault and resistance. Yet, as this passage unfolds, it moves progressively from a description of competitive violence to

2004), pp. 18–19.

80 Elizabeth Grosz, *Becoming Undone: Darwinian Reflections on Life, Politics, and Art* (Duke University Press, 2011), p. 46.

81 By 'a politics of identity' I mean here a form of political thinking constrained by a rigidly transcendental conception of identity as opposed to the term 'identity politics' which is frequently used to impugn any political thinking or movement that seeks to integrate or centre the voices, experiences, and theoretical *savoir faire* of marginalised, racialised, and other identities.

abstract affective states, and then to creative forms designed to produce and respond to desire. Indeed, by its conclusion, it even posits a caesura in the 'ever-acting' power of natural selection during which non-violence and withdrawal may prove just as 'successful' as their implied opposites.

Darwinian evolution, therefore, consists of two separate but interrelated processes of selection: a process of survival, according to which 'a grain in the balance will determine which individual shall live and which shall die'; and a process based on an organism's capacity to attract and be attractive. Grosz's focus on sexual selection rather than only on natural selection is in itself an important philosophical gesture, since it reframes what we understand by the signifier 'Darwin' as not only an avatar for 'the survival of the fittest' but also the name of a biologically grounded theorist of sexual relation, creativity, and reproductive possibility. For Grosz, moreover, Darwin is also the unlikely source of a feminist conception of biological reproduction. She reads Darwin in a Deleuzean mode, not only as a theorist of sexual difference but a theorist of the immanent possibility of transformation latent in evolution's historically contingent forms and their relations. The image of sexual selection offered by Darwin, therefore, may appear problematically heteronormative, but its focus on desire supplements the social-Darwinian principle of survival, derailing the instrumentality of bare life with relations defined by excess. 'The laws of sexual selection,' Grosz argues, 'are the principles of aesthetics, not the strategies of game theory; the functioning of appeal rather than the operations of rational agents who act according to their self-interests; the order of taste rather than [...] calculation'.[82] Such a reading of Darwin, then, allows us to read sexual selection as a process involved in the proliferation of biological and cultural life, in tension with the mechanics of natural selection.

So readers of Darwin — social theorists, philosophers, economists, and others — have produced radically divergent interpretations of his work. Evolutionary selection can either underwrite eugenics, free-market capitalism, and the sovereignty of the individual human subject or can fundamentally challenge the anthropocentrism latent in these ideologies, and the very ontological status of the human as such. Yet, both interpretive polarities in their way attest to a certain form of pessimistic optimism: the Spencerian conception of 'survival of the fittest' sees violence and domination as civilisation's condition of possibility and the reproduction of a higher form of human subject; the Deleuzean reading of the Darwinian-Copernican moment sees the dissolution of the human as we understand it as a predicate for the immanence and incalculability of political, social, and ontological transformation. For Howard Caygill, Deleuze fixates on the notion that natural selection ratifies the ontological primacy of 'difference in itself', such that his reading of Darwin ignores the brutality of the processes of selection through which difference is distributed.[83] This reflects the philosophical 'constructivism' in Deleuze's thinking that, it might

82 Grosz, *Becoming Undone*, p. 135.

83 Howard Caygill, 'The Topology of Selection: The Limits of Deleuze's Biophilosophy', in *Deleuze and Philosophy: The Difference Engineer*, ed. by Keith Ansell Pearson (Routledge, 1997), pp. 149–62.

be argued, is blind to the material effects of biological and historical forms of change.[84] Inverting these critiques, it might also be argued that Deleuze's Darwin, by privileging the proliferation of difference and not its consequences, suggests an image of being (as becoming) in which the violence of natural selection is but one part of a wider assemblage of processes of change, which include social, political, sexual, and aesthetic transformations. The social-Darwinian view, by contrast, is sanguine about the need for a robust system of selection (in the form of, say, free market competition or eugenic constraints on reproduction), and sees — or seeks to replicate — the deterministic dynamics of elimination and survival in all facets of life, beyond the 'purely' biological.

Through these divergent yet connected interpretations of Darwin, therefore, we arrive again at an impasse that mirrors the Freudian tension between civilisation and its animal other. By seeming to embrace the radical thesis of 'ineradicable animality', the social-Darwinian interpretation seems unable to construct a vision of the human which does not also repress that idea, while the Deleuzean view is willing to affirm the theoretical dissolution of the human, a practical exercise that too many concrete social-Darwinian political projects have themselves prosecuted in material terms.

Fundamentally, the thesis of this book is that the Darwinian novel — that is to say, the novels studied here, written consciously (and unconsciously) in response to the Darwinian decentring of the human — registers and examines this impasse, in a dialogical form that can be seen between and within the texts themselves. This book's first chapter on Zola demonstrates that the reception of Darwin in Zola's novels, *Le Ventre de Paris*, *Germinal*, and *L'Œuvre* allows us to see their depictions of Second Empire France as structured by natural selection in social, economic, and biological terms. Yet, these novels dramatise a more complex theoretical dialogue. On the one hand, they can be read to articulate an emancipatory and anti-deterministic conception of heredity, drawn from Deleuze's reading of Zola in his *Logique du sens* [*Logic of Sense*] (1969), and a conception of biological abjection, via Julia Kristeva, according to which the function of Zola's 'putrid' realism is to reinscribe a necessary division between self and other, or human and animal.

If Chapter 1 is concerned with Darwin's theory of natural selection and its literary and theoretical afterlife in the work of Zola, Chapter 2 focuses on Hardy's novels *A Pair of Blue Eyes* and *The Return of the Native* and their engagement with a Darwinian sexual selection through which we may elaborate a radical re-conception of human erotic relation and creativity that challenges naturalistic nihilism. Drawing on Grosz's reading of sexual selection, this chapter explores how these novels represent female sexuality and sexual courtship in ways that draw on Victorian and Darwinian conceptions of sexual difference, and which at the same time affirm excess, desire, and a conception of life that evades calculable logics.

Chapter 3 offers an analysis of three Utopian novels spanning the nineteenth, twentieth, and twenty-first centuries. Reading Samuel Butler's *Erewhon*, Aldous

84 Peter Hallward, *Out of This World: Deleuze and the Philosophy of Creation* (Verso, 2006), pp. 162–63.

Huxley's *Island*, and Michel Houellebecq's *La Possibilité d'une île,* I illustrate how each work can be read both as a response to Darwin and as a furtherance of the long tradition of Utopian literature inaugurated by Thomas More's *Utopia* (1516), as well as a satirical tradition to which Utopianism is historically and aesthetically linked. Drawing once more on the work of Deleuze, in his early essay on 'L'Île déserte' ['Desert Islands'], this chapter elucidates the various ways each of these novels dramatise in explicit ways the human longing to transcend evolutionary change, which is only present implicitly in the work of Zola and Hardy. In each novel, Darwinism provides the technological, biological, and political impetus for fantasies of biological withdrawal and transcendence, giving narrative form to an image of the internally divided human that Darwin's theory of evolution involves, and to which the concept of humanity offers so much resistance.

CHAPTER 1

❖

Émile Zola: The War of Nature

> As more individuals are produced than can possibly survive, there must in every case be a struggle for existence, either one individual with another of the same species, or with the individuals of distinct species, or with the physical conditions of life. It is the doctrine of Malthus applied with manifold force to the whole animal and vegetable kingdoms. — Charles Darwin (*Origin*: 34)

Writing in his biographical study of Émile Zola on the genesis of his Naturalist programme of novels, Armand Lanoux describes an encounter between Zola and Jules and Edmond de Goncourt. Meeting the Goncourts by invitation at their residence in Paris, Lanoux tells us that Zola, 'un homme inquiet, anxieux, profond, compliqué' [an uneasy, anxious, profound, and complicated man], admonished the brothers for seeming to neglect the importance of Darwinian evolutionary biology in their writing. Zola, Lanoux attests, made the following proclamation: 'Les caractères de nos personnages sont determinés par les organes génitaux. C'est de Darwin! La littérature, c'est ça!' [The constitution of our characters is determined by the genitals. That's from Darwin! That is literature!].[1] Unsupported in Lanoux's text by clearly referenced bibliographic source material, this reported outburst is nevertheless claimed to have taken place in 1868, soon after the publication of the first novel of the *Rougon-Macquart* series of novels and just three years after the publication of the Goncourts' most widely-known work, *Germinie Lacerteux*, in 1871. The provenance of Lanoux's account is unclear. Yet this anecdote is notable for the way it locates (or imagines) a key moment in the French reception of Darwin's work (however crudely understood) in the Parisian literary-intellectual milieu at the foundational point in the development of Naturalist forms of realist representation and Zola's *Rougon-Macquart* project.

Analysing the preface to the Goncourts' *Germinie Lacerteux*, Erich Auerbach describes the literary Naturalist programme of artistic purpose outlined there and comments on the character of its intellectual context:

1 Armand Lanoux, *Bonjour Monsieur Zola* (Hachette, 1962), p. 100. Lanoux alludes to and quotes from the *Journal Goncourt* in relation to this episode but no such outburst is evident in the relevant entry. Jean Jacques Yvorel also notes also notes a failure to locate Lanoux's source. See: Edmond de Goncourt and Jules de Goncourt, *Journal des Goncourts: mémoires de la vie littéraire*, 3 vols (Charpentier, 1887–88), III, pp. 21–23; Jean-Jacques Yvorel, 'L'Enfant des Halles', in *Les Halles: images d'un quartier*, ed. by Jean-Louis Robert and Myriam Tsikounas (La Sorbonne, 2004), pp. 107–18.

> The influence of the enthusiasm for science which marked the first decades of positivism, when all active intellects — insofar as they were consciously searching for new methods and values in accord with the times — strove to assimilate the experimental techniques of science.[2]

Zola's manifesto for the Naturalist novel, *Le Roman expérimental*, can be understood in the same context, as one of the more notable examples of this search for new scientific methods of cultural production. In keeping with Auerbach's observation, *Le Roman expérimental* is dedicated primarily to the elaboration of his project's debt to the work of the experimental physiologist Claude Bernard, and not to evolutionary biology. Nevertheless, Zola's essay does allude suggestively to the theoretical importance of Darwinian evolution in understanding the mechanisms of social and biological heredity. On the question of the influence of environment on human life, Zola admits: 'Il faudrait aborder les théories de Darwin; mais ceci n'est qu'une étude générale sur la méthode expérimentale appliquée au roman, et je me perdrais, si je voulais entrer dans les détails' [I ought to touch upon Darwin's theories; but this is only a general study of the experimental method as applied to the novel, and I should lose myself were I to enter into details].[3] As does Lanoux's anecdote, this oblique allusion to Darwinian science seems both to affirm and raise questions about the position of Darwin's work in Zola's imagination. Read chronologically, we might see the suggestive absence of engagement with Darwin in Zola's manifesto as that which makes Lanoux's later imaginative account of Zola's reception plausible. Read in reverse, if the story about Zola's meeting with the Goncourts seems to establish Zola's vulgar (in)comprehension of Darwin, it also, in turn, allows us to read the enigmatic brevity of Zola's reference to Darwin in *Le Roman experimental* as a (potentially false) kind of modesty, by contrast with the later reported outburst. Whether or not he had read Darwin's work, Zola seemed to understand enough to know that the question of Darwinian evolution demanded significant further engagement than was possible in the context of his statement of literary intent.

Did Zola read Darwin and integrate his thinking into the theoretical and practical project of literary Naturalism? Or does the questionable nature of Lanoux's anecdote, and Zola's allusion in *Le Roman expérimental*, suggest that the nature of Darwin's influence on Zola is less certain? As such, this story invites the same questions that Zola's fictional author Sandoz was concerned to ask in *L'Œuvre* about the relation of literature to science. What can the novelist extract from science and specifically Darwinian evolutionary theory? How and under what conditions can this extraction take place? As well as posing important questions about reception and transmission, this anecdote raises key theoretical questions about Naturalist literature's relation to Darwinian thought. By Lanoux's telling, it is from Darwin, at least in part, that Zola derives the notion that human 'constitution' and 'character'

2 Erich Auerbach, *Mimesis: The Representation of Reality in Western Literature*, trans. by Willard R. Trask (Princeton University Press, 2013), p. 496.

3 Émile Zola, *Le Roman expérimental*, 2nd edn (Charpentier, 1880), p. 18; *The Experimental Novel, and Other Essays*, trans. by Belle M. Sherman (Cassell, 1893), p. 19.

stem directly from a biological essence, located in the reproductive organs which stand, metonymically, for an evolutionary instinct, a will to survive and, therefore, for a biologically deterministic conception of human relations. As such, in this account, the Darwinian human is one wholly determined by what Freud might have called its 'ineradicable animality'. In what way does Zola's literary work reflect this conception of human determination? And to what extent can that reflection be understood as a way of rethinking the human in light of the Darwinian decentring of 'man'?

This chapter takes up these questions, examining a selection of Zola's Rougon-Macquart novels, *Germinal*, *L'Œuvre*, and *Le Ventre de Paris*, and how they respond to the Darwinian intervention in our conception of the human. While the version of Darwin that Zola is purported to espouse in Lanoux's biography seems largely concerned with genitality as a metonym for hereditary determination, this chapter focuses also on how in Zola's fiction hereditary determination works in concert with the Darwinian 'struggle for life' in determining the fate of the human subject. Reading Zola in this way reveals a complex dynamic at work. His works elaborate narratives driven by the agonistic relations of biological life in social and political terms according to which human life is determined by forces outside its control. Yet, the protagonists on whom these narratives focus — an artist and two political revolutionaries — are dedicated to transcending this condition through different forms of political and aesthetic agency. Moreover, Zola's references and allusions to Darwinian evolution in his novels, whether these are implicit or explicit, appear to thematise the very conditions under which Zola's fiction could call itself 'scientific'. As such, Zola's works seem both to reaffirm the crudely deterministic vision of life that Lanoux attributes to Darwin, while putting into question both its durability and the textual conditions of its reproduction.

The latter half of the chapter explores this dynamic further by turning to the work of Gilles Deleuze and Julia Kristeva, whose respective commentaries on Zola draw on psychoanalytic concepts of death drive and abjection to confront the narrative pessimism and representational violence of the *Rougon-Macquart* project. I engage both writers, together and against one another, to delve further into the psychic and discursive function of Zola's Naturalist narrative apparatus. Does Zolian Naturalism function to reproduce a reality defined by Darwinian determinism? Or can a renewed understanding of Zola's Darwin help us articulate a different kind of relation to evolutionary thought and the post-Darwinian human?

Darwin and Zola: Reception and Determinism

For Lanoux's Zola, a circular logic ensures the authority of the claim that truly realist literature should flow from a conception of biology. Its truth value as a theory is indexed by the self-evident prestige of the author ('C'est de Darwin!'), just as the author's prestige is derived from the truth of his theories. The task of the novelist, by this account, is less to examine the truth content of the underlying scientific knowledge but to elaborate a literature that is faithful to its scientific sources.

What are the historical and aesthetic implications of such a conception of post-

Darwinian literary production, according to which human life can be encased by biological conceptions of reproduction and survival? Commenting on the emergence of biological science and positivism in nineteenth-century Europe, José Ortega y Gasset links Zola's work to that of Darwin and a wider biological imaginary:

> Darwin believed he had succeeded in imprisoning life — our last hope — within physical necessity. Life is reduced to mere matter, physiology to mechanics. The human organism, which seemed an independent unity, capable of acting by itself, is placed in its physical environment like a figure in a tapestry. It is no longer the organism which moves but the environment which is moving through it. Our actions are no more than reactions. There is no freedom, no originality. To live is to adapt oneself; to drive us out of ourselves. Adaptation is submission and renunciation. Darwin sweeps heroes off the face of the earth.[4]

If early in the twentieth century, Max Weber's capitalist subject was locked in an 'iron cage' by social and economic determinants, 'bound to the technical and economic conditions' of the capitalist industrial mode of production, Ortega sees its proleptic biological analogue in the nineteenth-century subject of biological evolution. The forces of natural selection, like the 'irresistible force' of 'machine production' represent a totalising form of biological determination, the imprisonment of all human agency within a mechanical evolutionary logic of adaptation.[5] In response, Ortega suggests, literary realism of the nineteenth century itself sought to adapt to these new discursive and intellectual coordinates. Realism adapts, he suggests, echoing Levine's argument on the evolutionary narrative causality of the Victorian novel, by abandoning the kind of epic, heroic form in which the narrative's agential fulcrum lies with the actions of a protagonist. Instead, the realist novel takes up instead a narrative construct in which the individual is merely the residuum of supra-human social and biological forces.

In political and ethical terms, such resignation in the face of an anti-heroic biological backdrop defines both the realist author and the post-Darwinian subject (who in this account are one and the same). The realist human is a simulacrum of heroic sovereignty, deprived of authentic selfhood, resigned to its fate, incapable of resisting the constraints of evolutionary adaptation, renouncing the freedom that is thought to constitute human existence. The narrative apparatus of Zola's 'experimental novel', Ortega suggests, submits its subjects to the same life-denying fatalism, reaffirming degeneration, death, and dissolution, a construal of the predetermined relations and enforced adaptations of heredity and evolution. For Ortega and other major critics this narrative fatalism is reinforced by Zola's insistence on the overwhelming verisimilitude that infuses his works.[6] As the

4 José Ortega y Gasset, *Meditations on Quixote*, trans. by Evelyn Rugg and Diego Marín (University of Illinois Press, 1961), p. 164.

5 Max Weber, *The Protestant Ethic and the Spirit of Capitalism*, trans. by Talcott Parsons (Routledge, 2010), p. 123.

6 For Gabriel Josipovici, Zola's work could be categorised in negative terms as an example of 'extreme objectivity' in literature, seeking to conceal its own existence as a 'made object': Gabriel Josipovici, *The World and the Book: A Study of Modern Fiction* (Stanford University Press, 1971), pp.

human becomes indistinguishable from its environment, any scientific theory that can claim to represent relations which constitute the environment in its totality can by the same token lay claim to a panoptic representation of the human actor and its motivations, effectively depriving the human of its ontic autonomy on the literary and biological stage.[7] Michel Serres, using the vocabulary of a different scientific discipline, argues that the hereditary narratives of Zola's *Rougon-Macquart* cycle trigger a thermodynamic plot of entropic decline and tragic dissolution.[8] This view of Zola's representational determinism is articulated in the Marxian criticism of Gyorgy Lukacs, for whom Zola's Naturalism was an avatar of 'false' objectivity. Zola's use of a narrative mode of monological description, Lukacs argues, as opposed to the complex, multi-perspectival focalisation distinctive of Tolstoy or Balzac, presents the tragic fates of its characters as nothing more than the side-effects, the *caput mortuum*, of a predetermined process.

Yet, there is in these aesthetic and ethical assessments a sense of irresistible determination that conflates the discursive and the biological. There is the discursive force of the new knowledge of biology and evolution, expressed through the narrative medium of the novel (Zola is merely its conduit or its imitator), and there is the irresistible force of biological law itself, which is doubled, and thereby reproduced, in Zolian narrative and representation. The force of conceptual inheritance, it is suggested, is as strong as the force of heredity itself. For David Baguley, 'Naturalist literature unhesitatingly sees man as essentially formed by, and explained in terms of, the biological model: birth, life, decline, death, developing within the broader framework of evolution, destined to struggle and eventual waste'.[9] As such, nature, in Zola's work is characterised as 'ontologically evil', drawing humanity 'into an essential compliance with [its] laws [and abolishing] distinct humanity'.[10] What is at issue in Ortega's assessment, however, is the implication that this literary characterisation of nature, in its engagement or complicity with the ideology of biological determinism (as expressed in the formal rigidity of the Naturalist project), in fact socialises the apparent anti-humanism of the biological model of life.

Yet, it was Thomas Hardy who pointed out in 1891 that Zola's professed experimental method seems to 'reveal an obtuseness to the disproof conveyed in his own novels'.[11] Indicting Zola's fiction on grounds of being 'too realist' and too 'deterministic' merely takes at face value Zola's assertion of the scientific objectivity of his writing, while at the same time finding it inadequate. Similarly, Susan Harrow argues that the assumption that Zola's writing is too rigidly objective is

288–89. See also Henry James's assertion that Zolas's work was nothing more than the mimesis of scientific objectivity rather than objective in itself, in *The Art of Criticism: Henry James on the Theory and Practice of Criticism*, ed. by William Veeder and Susan M. Griffin (Chicago University Press, 1986), p. 446.

7 Ortega y Gasset, *Meditations on Quixote*, pp. 164–65.

8 Michel Serres, *Feux et Signaux de Brume, Zola* (Grasset, 1975).

9 David Baguley, *Naturalist Fiction: The Entropic Vision* (Cambridge University Press, 1990), p. 216.

10 Serres, *Feux et signaux de brume*.

11 Thomas Hardy, 'The Science of Fiction', in *Thomas Hardy's Public Voice*, ed. by Millgate, pp. 106–10 (p. 107).

contradicted by the 'mythopoeic evidence' in it.[12] For Henri Mitterand, Zola 'did not merely write a "natural and social history" of the Second Empire'; instead, the mythic, literary, and fabulist nature of his writing — its 'music of internal echoes and rhythms' — 'transcend purely realistic *mimesis*'.[13] And for Fredric Jameson, too, the narrative determinism of Zola's *Rougon-Macquart* project stands in tension with the novels' descriptive materiality, in which 'an excess of the sensory' works to enhance the 'gap between words and things'.[14] Each of these critiques index the idea that to criticise Zola's work for failing to be as objective as it seems to claim also suggests that a more suitably objective and accurately scientific approach to realism is possible or desirable. Moreover, by seeming to conflate, as Ortega does, the intellectual influence of nineteenth-century biology and the determinism of Zola's narrative rhetoric, we also ignore the complexity of the hereditary and evolutionary theories that influenced Zola, while also failing to examine the complex relationship of Zola's narratives and descriptive technique to those discourses.

A different kind of assessment of the relationship between Zola and Darwin might attend instead to the ambiguous evidence that marks Zola's reception of Darwin. The brief allusion to Darwin in *Le Roman expérimental* and its cryptic afterthought of course raises suspicion regarding the nature of Zola's erudition in Darwinian theory, a suspicion that should be deepened by the fact that remainder of that essay, and the remainder of Zola's writing career, passes without any detailed or explicitly systematic treatment of Darwinian science. If Zola's work, and its assumed pessimism, verisimilitudinal rigidity, and narrative determinism, was indeed influenced by Darwinian evolution, Zola certainly did not signal this influence in his notes or elsewhere.[15] Yet, the question might not be whether or to what extent Zola's literary Naturalist pessimism is derived inaccurately (or all-too accurately) from Darwin's evolutionary theory, but whether that 'reception' in the sense of a systematic and conscious engagement may not have taken place at all. Perhaps, then, a more historically and genetically appropriate account of Zola's scientific influences might include Claude Bernard's *Introduction à l'étude de la médecine expérimentale* (1865) which was integral to Zola's theory of literary scientific verisimilitude as he expresses it in in *Le Roman expérimental*. In this, Zola draws on Bernard's account of experimental medicine to support the notion that the novelist's

12 Susan Harrow, *Zola, the Body Modern: Pressures and Prospects of Representation* (Legenda, 2010), p. 94.

13 Henri Mitterand, *Émile Zola: Fiction and Modernity*, trans. by David Baguley (Émile Zola Society, 2000), pp. 3–6.

14 Fredric Jameson, *The Antinomies of Realism* (Verso, 2013), pp. 50–54.

15 This is supported by David Baguley's brief but excellent survey of Zola's reception of Darwin, which points out that, in addition to the allusion to Darwin in *Le Roman expérimental*, other references to Darwin appear only a handful of times across Zola's entire fictional corpus. Accepting that Zola, as a novelist, may have preferred to integrate or allude to Darwinian images, metaphors, ideas, or signifiers into the texture of the novel, Baguley also points out that very little evidence can be found — bar a selection of brief notes or allusions — of Zola's engagement with Darwin in the detailed *ébauches* that formed the evidentiary basis for Zola's observational scientific literary project. David Baguley, 'Zola and Darwin: A Reassessment', in *The Evolution of Literature: Legacies of Darwin in European Cultures*, ed. by Nicholas Saul and Simon J. James (Rodopi, 2011), pp. 201–12.

'observation should be an exact representation of nature' and that, in turn, the literary *corpus* might act as a space in which to conduct sociological experiments and develop scientific theses about nature and humanity.[16] A study of Zola's theory of heredity might examine the work of the French doctor and theorist Prosper Lucas whose work in *Traité philosophique et physiologique de l'hérédité* (1847–50) was instrumental in Zola's conception of the genealogical tree structure upon which his *Rougon-Macquart* cycle of works is based.[17] Further analysis might include French translations of Cesare Lombroso's *L'uomo deliquente* (1876) and its thesis on the atavistic origins of human criminality which is reflected in Zola's literary studies of criminal pathology.[18]

Why, then, focus on Darwin in Zola's work at all? Perhaps, as Gillian Beer suggests, Zola's fiction was influenced by a wider cultural shift in which Darwinian ideas had transformed into unconscious assumptions. Yet, although it is true that there is a dearth of systematic references to Darwin in Zola's writings at large, and only a handful of references in Zola's fiction, these references are more significant for precisely that reason. Rather than assume Zola was either a systematic Darwinian or wholly ignorant of Darwin's work (and 'merely' using the name Darwin as a pretence), these fictional allusions invite us to attend to ways in which Zola uses the ideas, images, and signifiers associated with Darwin (however unsupported by systematic reading) for the purposes of his fiction.

For Baguley, whose survey of Zola's reception of Darwin looks across the twenty works of the *Rougon-Macquart* series, as well as later works (The *Les Trois Villes* trilogy, 1893–98, and the unfinished *Les Quatres Évangiles* tetralogy, 1898–1902), Zola's fiction can be read to articulate a coherent vision of Darwinian theory that is specifically social and political in nature.[19] *Au Bonheur des Dames* (1887), he notes, dramatises a vicious economic 'struggle for life' between competing Parisian department stores. *Germinal* (1885), too, he writes, offers a vision of social and industrial strife framed explicitly in Darwinian terms of competition and extinction, while *La Débâcle* (1892), and in it Zola's fictional account of the Battle of Sedan, was framed as a specifically Darwinian (and apocalyptic) form of political and violent struggle. These socio-politically grounded readings of Zola's Darwin are consonant with the wider context of the reception of Darwin in nineteenth-century France, which was characterised by some significant resistance, not only on the part of anti-evolutionist conservatives but also by liberals and biologists who remained committed to Lamarckian mechanisms of change. This dynamic can be seen in the first translation of Darwin's *On the Origin of Species*, by Clémence Royer,

16 Zola, *The Experimental Novel, and Other Essays*, trans. by Sherman, p. 7.

17 David Baguley, 'Darwin, Zola and Dr Prosper Lucas's "Treatise on Natural Heredity"', in *The Literary and Cultural Reception of Charles Darwin in Europe*, ed. by Glick and Shaffer, III, pp. 416–31. For an account of the importance of Lucas to early hereditary science see Carlos López-Beltrán, 'In the Cradle of Heredity: French Physicians and *L'Hérédité Naturelle* in the Early 19th Century', *Journal of the History of Biology*, 37.1 (2004), pp. 39–72, doi:10.1023/b:hist.0000020385.44164.e2.

18 Marie-Christine Leps, *Apprehending the Criminal: The Production of Deviance in Nineteenth Century Discourse* (Duke University Press, [1992]), pp. 166–76.

19 Baguley, 'Zola and Darwin'.

which is prefaced by a lengthy anti-clerical essay commenting on the significance of the theory of natural selection and 'the struggle for life':

> Il résulte de cette lutte universelle un choix, une sélection naturelle (natural selection) constante des races et des individus les mieux adaptés aux circonstances de temps et de lieu; de sorte que les êtres les plus parfaits, relativement à ces circonstances, l'emportent sur les êtres les moins parfaits qu'ils tendent à supplanter et à détruire, si ces derniers ne trouvent pas le moyen d'émigrer.[20]
>
> [Out of this universal struggle a choice is made, a constantly recurring natural selection of the races and the individuals best adapted to the circumstances of their time and place, such that the most perfect beings, relative to those circumstances, prevail over those less perfect beings, whom they tend to supplant and destroy, if the latter cannot find the means to emigrate.]

Royer, liberal and left-wing social thinker, explicitly politicises Darwin's theory of natural selection, accepting the Malthusian basis of the struggle for life ('Les hommes sont inégaux par nature' [Men are unequal by nature]), praising Darwin's work in *On the Origin of Species* as 'la revelation rationelle du progrès' [the rational revelation of progress], while anticipating eugenic notions of social improvement.

Read this way, it is possible to see Royer's framing of Darwin's work as implicitly Lamarckian in its privileging of progressive evolutionary teleology, as well as social-Darwinian in the way it places faith in the process of natural selection to eliminate weakness in the social organism.[21] Arguably, just as important in the dissemination of Darwin in nineteenth-century France was the Barré-Lebiez affair of 1878, during which it emerged that the perpetrator of a highly-publicised murder was a committed social-Darwinian, a connection that was instrumentalised by conservative opposition to Darwinian theory more widely. The outcome of this attempt to discredit Darwinism was an enhanced recognition of Darwinism in France, but in a way that privileged less the 'theory itself in the abstract' and focused instead on 'the array of discursive positions taken up in response to it, and how these positions enabled the construction of rival epistemological, social, and political identities'.[22] In the years following this controversy, it is argued, French republicanism was ideologically indebted to a 'Darwinian culture', and Darwinism 'became the center of debates on ethical and moral principles in society'.[23] The significance of the specifically political and social conception of Darwin in nineteenth-century France is underlined by the conflict between the 'struggleforlifeurs', the bourgeois proponents of scientific naturalism, and their conservative opponents. The former

20 Clémence Royer, 'Préface de la première édition', in Charles Darwin, *De l'origine des espèces par sélection naturelle, ou des lois de transformation des êtres organisés*, ed. and trans. by Clémence Royer, 3rd edn (Guillaumin, 1870), pp. xxvii–lxxi (p. lxii); my translation.

21 Joy Dorothy Harvey, *Almost a Man of Genius: Clémence Royer, Feminism, and Nineteenth-Century Science* (Rutgers University Press, 1997), pp. 210–11; Michel Prum, 'Charles Darwin's First French Translations', in *The Literary and Cultural Reception of Charles Darwin in Europe*, ed. by Thomas F. Glick and Elinor Shaffer, 4 vols (Bloomsbury, 2014), iv, pp. 391–99.

22 Liv Grjebine, 'A Darwinian Murder: The Role of the Barré-Lebiez Affair in the Diffusion of Darwinism in Nineteenth-Century France', *Isis*, 113.4 (2022), pp. 689–709 (p. 707), doi:10.1086/722539.

23 Jean-Marc Bernardini, *Le Darwinisme social en France (1859–1918): fascination et rejet d'une idéologie* (CNRS, 1997), p. 99; Grjebine, 'A Darwinian Murder', p. 708.

group, like Royer, saw in the dynamics of the 'survival of the fittest' a mechanism that would dissolve entrenched hierarchies and could ensure social progress, an idea to which anti-republican intellectuals were vehemently opposed. Yet, in their own way these conservative opponents were as much proponents of a biologically informed conception of society as the *struggleforlifeurs*. The naturally competitive character of the evolutionary state of nature, for them, demanded instead a specific kind of 'civilisation': a conservative, anti-democratic, authoritarian state form that would repress such savage instincts.[24]

It is via the prism of this politicisation — both by conservative and progressive ideological camps — that Zola is likely to have encountered Darwin, and even the brief overview of the novels offered by Baguley seems to confirm this. For even if we lack evidence that Zola did systematically engage with Darwinian ideas, it is also clear that Darwin's name, as a signifier for a set of biological and political ideas about transformation, progress, and struggle, would have formed part of Zola's imaginary. The following section, therefore, examines Zola's Darwinism in one of his most canonical novels, *Germinal*, to explore the way in which Royer's 'lutte universelle' was reproduced as well as questioned in Zola's work. Contrary to the view of Zola as either a naïve or malign Darwinian, captured by the biological pessimism of his age, his fiction instantiates a keen awareness both of the political and social implications of the social-Darwinian worldview (in both its pessimistic and optimistic forms) as well as the complex and problematic discursive conditions from which these conceptions had emerged.

The Struggle for Life: Germinal

Germinal (1885), perhaps Zola's most canonical and widely recognised novel, was conceived shortly after the events of the Paris Commune in 1872 and was nourished by two decades of social and political upheaval in France that included the birth of an organised class-conscious proletariat, waves of strikes and repression (including violent strike action at the coalmines of Montceau-les-Mines), and the efflorescence of socialist thought in Europe.[25] The novel (Zola's second *roman ouvrier* after *L'Assomoir*) depicts the social and political struggle of the inhabitants and workers of a small fictional mining town in the north of France, suggestively named 'Montsou' [Penny Mountain]. The coal mines of Montsou represent the community's sole means of existence, and the novel details with a characteristic focus on technical detail the conditions of the workers' exploitation, the emergence of a revolutionary consciousness and instigation of a widespread strike, and its eventual collapse in

24 Louise Lyle, '"Le Struggleforlife": Contesting Balzac through Darwin in Zola, Bourget, and Barrès', *Nineteenth-Century French Studies*, 36.3–4 (2008), pp. 305–19. These *struggleforlifeurs*, Lyle demonstrates, were the subject of a series of conservative anti-Republican, late nineteenth-century *romans à thèses* which, ironically, criticised their social application of Darwinism as 'a mechanism which serves to uphold the privileges of the economically and politically dominant few at the expense of the impoverished and disenfranchised many' (p. 315).

25 For a useful introduction to the context and genesis of the novel see Henri Mitterand's 'Introduction' (*Germinal*: 239–46, esp. 239).

the face of impossible deprivation, intransigent management, and the unassailable power of capital.

An indication of Zola's biological framing of political revolution can be found in his preparatory *ébauches* for the *Rougon-Macquart*, in which, according to Colette Becker, we find an early sketch for the novel's protagonist, Étienne Lantier, the putative leader of the strike action. In its first conception, in the *ébauche* of 1884, Étienne was conceived not as the heroic instigator of a labour strike as he is in the finished work, but as a murderer ('maniaque de l'assassinat dans la famille' [a maniacal killer in the family]) driven by biological instinct. This characterisation is anticipated in the notebooks of 1868, when Zola outlines his vision for Jacques Lantier, the murderous protagonist of *La Bête humaine*: 'Un des cas étranges de criminal par hérédité qui, sans être fou, tue un jour dans une crise morbide, poussé par un instinct de bête' [One of those strange cases of a hereditary criminal who, without being mad, one day kills in the midst of a morbid crisis, driven by animal instinct].[26] Of course, while Étienne's character is developed in a different direction, it should not escape our notice that Jacques and Étienne share a hereditary connection as brothers, and that both are transgressive figures in their own way — one politically, the other morally. Additionally, Zola's notes also hint at a biologically pessimistic understanding of the 'sauvagerie' of revolution and its violent consequences.[27] There is a link then, in this early conception of the novel, between a positive conception of the worker and revolutionary consciousness and the so-called 'lesion' of tragic biological inheritance, which associates political disruption with criminality.

The novel examines this tension in explicit detail as the narrative draws to its conclusion. Following a massacre of fourteen strikers and the dawning realisation that the strike action is likely to collapse, Étienne, and his comrade Souvarine, two of the movement's key actors, debate the merits and demerits of Darwin's theory of natural selection:

> Il se faisait une idée révolutionnaire du combat pour l'existence, les maigres mangeant les gras, le peuple fort dévorant la blême bourgeoisie. Mais Souvarine s'emporta, se répandit sur la bêtise des socialistes qui acceptent Darwin, cet apôtre de l'inégalité scientifique, dont la fameuse sélection n'était bonne que pour des philosophes aristocrates. (*Germinal*: 508)
>
> [He formed a revolutionary idea of the fight for existence, the lean swallowing the fat, the strong people devouring the sickly bourgeoisie. But Souvarine got angry and held forth about the stupidity of the socialists who accepted Darwin, that apostle of scientific inequality, whose famous natural selection was only fit for aristocratic philosophers.] (*Germinal*: 450)

Here, Darwinian evolution is framed in relation to its value for the practical

26 Colette Becker, 'Du meurtrier par hérédité au héros révolutionnaire: Étienne Lantier dans le dossier préparatoire de *Germinal*' <http://www.item.ens.fr/articles-en-ligne/> [accessed 19 July 2024]; see also Colette Becker, *Émile Zola, la fabrique de Germinal: dossier préparatoire de l'œuvre* (SEDES, 1986).

27 Becker, *Émile Zola, la fabrique de Germinal*, p. 258.

and theoretical project of proletarian emancipation. Étienne, echoing Royer, remains optimistic despite having witnessed the death of his comrades at the hands of the state, and reads the struggle of natural selection as the scientific basis for political progress. In opposition stands his comrade, Souvarine, who rages against Étienne's conflation of biological and historical determinism, offering a Marxian *Ideologiekritik* of the capitalist economic undertow of the idea of natural selection. Darwin, Souvarine argues, in an echo of Marx's letter to Engels cited in the Introduction, is nothing other than an 'apostle' of inequality, the purveyor of a theory that naturalises competition, inequality, and suffering, fit only for 'aristocratic philosophers'. Anyway, Souvarine argues, if indeed the biological struggle for life governs agonistic relations between classes, would not that very mechanism go on to threaten the proletariat when it assumes power?

> N'était-il pas à craindre que le monde nouveau ne repoussât gâté lentement des mêmes injustices, les uns malades et les autres gaillards, les uns plus adroits, plus intelligents, s'engraissant de tout, et les autres imbéciles et paresseux, redevenant des esclaves? Alors, devant cette vision de l'éternelle misère, le machineur cria d'une voix farouche que, si le justice n'était pas possible avec l'homme, il fallait que l'homme disparût. Autant de sociétés pourries, autant de massacres, jusqu'á l'extermination du dernier être. (*Germinal*: 508)

> [Wouldn't there be a danger that the new world might grow up gradually spoiled by the same injustices, some people sick and others healthy, some more skilful and intelligent, succeeding in every venture, others stupid and lazy, becoming slaves again? And then, faced with this vision of eternal misery, the mechanic cried out in a ferocious voice that if justice was impossible in a world of men, then mankind would have to disappear. For every corrupt society there should be another massacre, until the last human being had been exterminated.] (*Germinal*: 450)

Although notionally opposed to one another, both of these interpretations of Darwin share an investment in a certain kind of biological and social determinism. In its humanistic socialist form, for Étienne, the biological superiority of the proletariat underwrites the historical inevitability of socialist emancipation — a human society that frees itself from the struggle of social and biological change. In its negative form, Souvarine points out, the identification of the proletariat with the dominant evolutionary power is an arbitrary one; the only guarantee made by the Darwinian law of struggle is the constancy of war and the eventual extermination of the human as such. If heredity provides the structural conceit for the *Rougon-Macquart* project, here a notionally Darwinian conception of struggle characterises human relations within that framework, according to which the human can either transcend or be wholly exterminated by evolution.

Yet, the formally dialogic nature of Zola's literary interpellation of Darwin's theory complicates any effort to define Zola's work, in this novel and more widely, in this simplistic way. Counterposing Étienne's fetishisation of biological struggle and Souvarine's nihilistic resignation presents Darwin's theory of natural selection not as a theoretical fait accompli, but an object of contested interpretation, mediated by ideological conviction (and in Souvarine's case, a pessimism derived from the personal trauma of his wife's death). As such, this short but critical section in *Germinal* undermines the idea that Zola's work is naïvely dedicated to an impossible realist objectivity grounded in a faulty interpretation of natural science. Indeed, we are invited to view each of these artlessly scientistic interpreters of Darwin critically, even as the theoretical dilemma they outline is a genuine one. As such, the dialogue here resists monological or presumptively authoritative versions of Darwin in favour of a self-consciously critical presentation that reveals the internal tensions of its political implications.

This short section in the novel also indexes wider historical considerations in relation to the intersection of Darwinian natural selection and its social, political valences. David Stack in his history of socialism and Darwinian theory writes that:

> On the one hand, by ensuring the acceptance of evolution it opened up [for the left] the possibility for change. On the other, by explaining evolution in terms of natural selection it seemed to simultaneously undercut radical and socialist politics, as [...] smooth teleological progress [...] gave way to Malthusian brutality and wastefulness.[28]

Souvarine's and Étienne's arguments stage the ambivalence that also characterised the socialist and Marxist view of Darwin in their time. In 1862, Marx was critical of natural selection's ideological subtext ('with its division of labour, competition, opening-up of new markets, "inventions", and the Malthusian "struggle for existence"') yet only a year earlier affirmed the fundamental materialism of Darwin's approach in a letter to Ferdinand Lassalle: 'Darwin's book is very important and it suits me well that it supports the class struggle in history from the point of view of natural science'.[29] In this earlier letter, however, Marx's interpretation diverges from the evolutionary socialisms of either Souvarine or Étienne, in arguing that natural selection 'deals the death-blow to "teleology" in the natural sciences', underlining the fundamental materialism of Darwin's approach as against the vulgar historical determinism it is supposed to support. Engels, like Marx, was similarly critical of the bourgeois character of Darwin's theory of natural selection. In 'The Part Played by Labour in the Transition from Ape to Man', Engels wrote that:

> Even the most materialistic natural scientists of the Darwinian school are still unable to form any clear idea of the origin of man, because under this ideological influence they do not recognise the part that has been played therein by labour.[30]

Yet, also like Marx, Engels saw value in the materialism of Darwinian thought, avowing his acceptance of the Darwinian conception of human evolution, but regarding the 'method of proof (struggle for life, natural selection)' as a necessarily 'provisional' approach to the novelty of Darwin's discoveries.[31] For both Marx

28 D. A. Stack, 'The First Darwinian Left: Radical and Socialist Responses to Darwin, 1859–1914', *History of Political Thought*, 21.4 (2000), pp. 682–710 (p. 689).

29 Karl Marx, letter to Engels, 18 June 1862, in *Karl Marx and Friedrich Engels: Selected Correspondence*, ed. by Ryazanskaya, p. 128; Karl Marx, letter to Ferdinand Lassalle, 16 January 1861 (ibid., p. 115).

30 Friedrich Engels, 'The Part Played by Labour in the Transition from Ape to Man', in *Dialectics of Nature*, trans. by Clemens Dutt, 2nd rev. edn (Progress, 1986), pp. 170–83 (p. 178).

31 Friedrich Engels, letter to Pyotr Lavrovich Lavrov, 12[–17] November 1875, in *Karl Marx and*

and Engels, then, Darwinian evolution (with special reference to natural selection) represented a double-edged body of thought. On the one hand, it offered to put Marxist theory on a more materialist, anti-metaphysical footing; on the other, it appeared to provide a natural basis for the capitalist fetish for competition they both sought to oppose.

If Étienne and Souvarine's dialogue in *Germinal* reflects the contradictions inherent in a political co-option of Darwinian natural selection, it may also illuminate the ambivalent function of Zola's fiction in relation to questions of revolutionary political change. For while his works are often critical of the complacency, greed, and shallowness that in his view characterised the Second Empire bourgeoisie, Zola himself is said to have remained sceptical of mass, class-based action.[32] The role of Zola's novel cycle, in theory, had an important role to play in offering fictional experiments capable of contributing to the improvement of the social conditions of the people and places on which his works fixed their observational gaze. The dialogue between Souvarine and Étienne stages the theoretical impasse produced of any attempt to fuse a foundationally deterministic Darwin with social progress. Far from being straightforwardly compliant with the life-denying socio-biological determinism of the nineteenth century, Zola's work here presents Darwin's work as a discursive object open to theoretical contestation.

In so doing, this dialogue also offers a metacommentary on the reception context of its own production. Étienne, the narrator informs us, 'en avait lu des fragments, résumés et vulgarisés dans un volume à cinq sous; et de cette lecture mal comprise' [had read some fragments summarized and popularised in a five-sou volume; and from this half-digested reading] had developed his revolutionary conception of evolution (*Germinal*: 508; 450). This narrative aside indexes the reception of Darwin in France and in other European contexts in the latter half of the nineteenth century, which was often characterised by misinterpretation, through social theory, editorialisation, translations, and transformations, of which Royer's preface to her translation of *On the Origin of Species* is emblematic. This characterisation of Étienne's reading functions to ridicule not only the character's intellectual pretensions, but also those of 'darwinisant' intellectuals more widely, and of course Zola himself.[33] By staging the act of interpretation itself (and demonstrating a knowledge of the kinds of interpretations that were common at that time), the novel anticipates criticisms that argue Zola's Naturalism in practice represents a misguided scientism. The irony of Zola's implicit autocriticism is that it still does not allow us to reimagine him as a sophisticated thinker or reader of Darwinian thought but reveals something more complex instead. It underlines the discursive and historical

Friedrich Engels: Selected Correspondence, ed. by Ryazanskaya, p. 283.

32 Instead, Brian Nelson argues, Zola's political disposition was 'technocratic' and largely conservative, interested in 'efficient leadership and the rational organization of society' brought about by empirical study, a view that is in some ways the dominant one of our own political era: Brian Nelson, 'Zola and the Ideology of Messianism', *Orbis Litterarum*, 37.1 (1982), pp. 70–82 (p. 71), doi:10.1111/j.1600-0730.1982.tb00790.x.

33 David Baguley uses the term 'darwinisant' in relation to Zola to signify his complex relation to Darwinian thought: 'very much in the spirit of Darwin's heritage, never a disciple': Baguley, 'Zola and Darwin', p. 212.

contingency of Zola's notional (failed) objectivity, implicitly acknowledging a kind of ignorance on Zola's part, precisely at the point where his work is notionally at its most scientifically engaged. We may read the novel not as naïvely objective or inaccurately Darwinian and instead as engaging with its own apparent failure of erudition.

In addressing its discursive genesis, this section also thematises the question of the possibility of human existence and agency after Darwin. If in this post-Darwinian representation of Second Empire France, social and biological life comes to be defined by constant struggle, Étienne and Souvarine respond to that predicament by offering two different prospects for human liberation: nihilistic resignation to its impossibility, or action on the basis of delusional optimism. Yet, the dialogue ends by casting these solutions in irresolution. If it is sometimes thought that Zola's fiction functions not to liberate the human subject but to reconfirm a dreary biological determinism; the impasse at which Étienne and Souvarine arrive suggests a more complex picture. This is achieved not, as the apocalyptic romanticism of Ortega would have it, by simply lamenting the advent of Darwinian science and its implications for human existence (which Freud might suggest is symptomatic of the very same defensive disposition that characterises a more straightforward anti-Darwinism). Rather, this dialogue on Darwin dramatises an interpretation of his natural selection in context, as theory in action, inviting us to see that the influence of Darwin was neither monologically pessimistic (either in Zola's fiction or in the wider climate of nineteenth-century European thought), nor axiomatic in its influence.

This moment of discursive and theoretical aporia is echoed by the narrative in the novel's conclusion. Having tracked the development of a widespread strike, its social consequences, and its violent collapse, the novel closes with the image of Étienne leaving Montsou for Paris, filled with renewed revolutionary energy. The landscape Étienne traverses is littered with the corpses of his failed revolution: days prior to this, Souvarine sabotages a mineshaft, killing himself and many others, in a self-immolating response to the capitulation of the strike under pressure from the state and management. Étienne is struck by the suggestion that Souvarine's success in a violent instantiation of a nihilistic critique of evolutionary progress invalidates his own teleological and Marxian reading of natural selection: 'Darwin avait-il donc raison, le monde ne serait-il qu'une Bataille, les forts mangeant les faibles, pour la beauté et la continuité de l'espèce?' [Was Darwin right then, was the world nothing but a battlefield where the strong ate the weak, for the beauty and survival of the species?] (*Germinal*: 551; 521). Yet, his conclusion is as certain as before: 'Du sang nouveau ferait la société nouvelle. Et dans cette attente d'un envahissement des barbares, régénérant le vieilles nations caduques, raparaissait sa foi absolu à une revolution prochaine' [New blood would create a new society. And in his expectation of a barbarian invasion which would regenerate the decadent old nations there appeared his absolute faith in a forthcoming revolution] (551; 521). The narrative's conjunction of Étienne's blind Darwinian optimism and the consequences of Souvarine's murderous anti-humanism enacts a formally distinct but structurally analogous type of irresolution to that offered by the earlier discursive stalemate. By demonstrating the catastrophic consequences of Souvarine's and Étienne's respective commitments to a *Darwinisant* revolutionary socialism or nihilism, the novel underlines the contingency of the violent collapse of the workers' struggle. Étienne's intransigent insistence on the continuation of the strike and its fatal consequences, Souvarine's suicidal commitment to violence and his sabotage of the mine: these actions in the novel stem not from biologically essential traits or dispositions but instead can be understood to derive from these characters' respective commitments to a belief in biological determinism.

Germinal offers us an unexpectedly constructivist meditation, therefore, on the relationship between the reception of biological evolutionary theory and social, biological determination, according to which (mis-)reading and interpretation of Darwin are the contingent drivers of — and not merely responses to — concrete social, political, and economic struggles. This suggests, too, that if Darwinian determination is in that sense discursively constructed and historically contingent, a different way of reading might offer us and the novel's revolutionaries a way out of the tragic bind that the social-Darwinian account of natural selection seems to effectuate. The novel, however, offers no clear outline of a new theoretical or exegetical relation to Darwin's work that might support a different set of social and political actions, consequences, or theoretical possibilities. Yet, the very existence of this dialogue on Darwin in the novel at least suggests the possibility of a specific (if limited) kind of human agency. Contrary to an image of complicity in the face of an irresistible Darwinian influence, it reveals the fundamentally discursive nature of Darwinian concepts, establishing through the instantiation of critical contemplation a discursive interval, a caesura of disagreement, in the apparently irresistible, causal logic of Darwinian thought.

The Weak and The Strong in *Le Ventre de Paris* and *L'Œuvre*

To what extent is this dynamic present in the other two novels with which this chapter is concerned? As in *Germinal*, these works articulate a dialectic between revolutionary possibility and the collapse of revolutionary hope through hereditary determination and social-Darwinian struggle. *Le Ventre de Paris* (1873) is the third novel in the *Rougon-Macquart* series, and centres on the narrative of Florent Quenu. Following his illicit return to Paris after an exile to the Île du Diable [Devil's Island] (a proxy for the French penal colony of Cayenne in French Guiana), in the wake of Louis-Napoleon's 1851 *coup d'état*, the novel details his ensuing struggle to survive in the viciously competitive political and biological milieu of Second Empire Paris, as symbolised by the food markets of Les Halles. Florent's narrative is tragic. He is reunited with his brother's family, the comfortably petit-bourgeois owners of a *charcuterie*. Despite securing employment as a market inspector in Les Halles, and maintaining the pretence of bourgeois respectability by lodging with his brother and his wife, Florent's involvement with radical political circles sees him betrayed by his sister-in-law, arrested, and exiled once more.

We can read the novel as a critique of a conservative and decadent bourgeoisie, obsessed with consumption and self-interest, symbolised by a culture of complacent epicureanism and avarice at the centre of which Florent, a political fugitive and putative radical, finds himself. Yet Florent's first and second exiles from Paris call to mind Royer's view that those who fail in the struggle, those less well-adapted, might emigrate, if they are not killed. Anticipating the framing of evolutionary struggle in the later *Germinal*, this echo of social-Darwinian selection is developed in the novel in its thematisation of an inherent 'bataille des Gras et des Maigres' [battle between the fat and the thin]. Quenu, returning to Paris from exile, is educated by the artist Claude Lantier (the brother of Étienne and Jacques) who explains the originary myth of their shared social milieu: 'Cain était un Gras et Abel un Maigre. Depuis le premier meurtre, ce sont toujours les grosses faims qui ont sucé le sang des petits mangeurs... C'est une continuelle ripaille, du plus faible au plus fort, chacun avalant son Voisin et se trouvant avalé a son tour' ['Cain,' explains Claude, 'was a Fat man and Abel a Thin one. Ever since that first murder, the big eaters have sucked the lifeblood out of the small eaters. The strong constantly prey on the weak; each one swallows his neighbour and then gets swallowed up in turn'] (*Le Ventre*: 398; *Belly*: 191). As opposed to a Freudian conception of the originary sacrifice of a primal father that inaugurates societal prohibitions, this narrative of 'the story of Cain and Abel as retold by Darwin' instead elaborates a society according to which morality is subordinate to biological necessity and consumption.[34] The novel, therefore, seems to invite us to read Florent's Paris, which on his return from exile he finds 'gras, superbe, débordant de nourriture' [fat and sleek, overflowing with food] (254; 12), as a stage in which social relations are imbricated with biologically determined hierarchies, symbolised by the dichotomy of the fat and the thin, the well-adapted and the mal-adapted.

Florent, therefore, for whom 'le faim ne l'avait plus quitté' [hunger had never left him] (254; 11), is not merely poor and a political suspect on his return to Paris but is also marked — like the Lantier brothers — as a member of the hereditary class of the thin. His constitution is described by the narrator as 'le maigre, le profil noir et louche' [thin, sickly — suspect] (316; 89). He is subsequently confronted by a world of comparatively 'healthy' bourgeois power, as exemplified by his sister-in-law, 'La Belle Lisa', 'avec sa forte encolure, ses lignes rondes, sa gorge qui avançait, mettait une effigie de reine empâtée' [with her thick neck, rounded hips, and swelling bosom, looked like the queen of all this dangling fat] (295; 63). In that light, we might read Florent's eventual (but amateurish) plans to instigate a political insurrection not only as a political or domestic rebellion (against the bourgeoisie and his own family) but as a revolt against his biologically determined constitution and the social dynamics of competition. Florent's second exile is predictable from the hereditary and evolutionary logic of the novel. For Florent, his weakness (and, therefore, his failure) is determined by his heredity, which consigns him to fail in the struggle for life against the biologically well-adapted, well-fed, and contented bourgeoisie.

34 Naomi Schor, *Zola's Crowds* (Johns Hopkins University Press, 1978), p. 27.

Zola's later novel, *L'Œuvre* (1885–86), the fourteenth in the series, displaces the domain of universal struggle to the art market of a fictional Parisian Impressionist milieu. The novel, which draws on Zola's personal relationship with Paul Cézanne, on whom the novel's protagonist is modelled, depicts through the story of the painter Claude Lantier (whom we met first in *Le Ventre de Paris*) a cut-throat milieu of Parisian artists and intellectuals. Claude's narrative is concerned both with his economic and professional frustrations (his attempts to have paintings accepted at Paris Salons) as well as the more profound failure to produce work that satisfies the *avant garde* aesthetic and ambition, the reason his work fails both critically and financially. Significantly, Claude's pessimistic diagnosis of nature and society in *Le Ventre de Paris* — the ouroboros of the strong devouring the weak — proleptically links the novel's plot with the narrative of Darwinian struggle as well as the question of hereditary weakness. For Claude, the obstacle to his success, apart from the academic conservatism of the Salons and the superior commercial acuity of his competitors, is the nature of his own obsessional but undeveloped genius, which is framed in the novel as a hereditary fault. Claude was inhibited, the narrator tells us, by 'le désequilibrement des nerfs dont il souffrait, le détraquement héréditaire qui, pour quelques grammes de substance en plus ou en moins, au lieu de faire un grand homme, allait faire on fou' [some maladjustment of the nerve centres, or through some hereditary flaw, which because of a gram or two of substance too much or too little, instead of making him a great man was going to make him a madman] (*L'Œuvre*: 179; *Masterpiece*: 282). If for Darwin, in *On the Origin of Species*, a 'grain in the balance' represents the distinction between survival and extinction, for Claude a 'gram' of an unnamed biological substance distinguishes between success and madness. The phenomenology of this knowledge of hereditary fate, for Claude, is that of frustration and hopeless despair in the face of supra-human forces:

> Il s'affolait davantage, en s'irritant de cet inconnu héréditaire, qui parfois lui rendait la création si heureuse, et qui d'autres fois l'abêtissait de stérilité, au point qu'il oubliait les premiers éléments du dessin. Et sentir son être tourner dans une nausée de vertige. (*L'Œuvre*: 49)

> [What drove him to distraction was the infuriating thought of the hereditary something, he did not know what, that sometimes made creation a sheer pleasure and other times reduced him to such complete sterility that he forgot the very basics of drawing. It was like being swept up in a sickening vortex.] (*Masterpiece*: 45)

The ambiguity of the free-indirect style used here suspends an attempt to determine whether the 'hereditary flaw' is a self-created product of Claude's own obsessionality or the proper symptom of a thoroughgoing biological determinism related objectively by an uninvolved narrative voice. Whatever the aetiology of his condition, biological or psychical, its effect is the same: throughout the novel, he experiences repeated frustration at his inability to overcome his own obsessional desire for perfection.

Claude's attempt to create original art is in effect, therefore, an attempt to transcend his biological condition, as he understands it and as it is framed by the novel's narrative (and the wider narrative schema of the *Rougon-Macquart* series). It also echoes Florent's plans to resist the imprisonment of socio-biological predestination. In his preparatory notes for *L'Œuvre*, Zola describes wanting to represent 'la lutte de l'artiste contre la nature' [the struggle of the artist against nature], registering Claude's aesthetic battle against biology: the struggle to paint and draw what stands in front of him and, in so doing, to dominate nature via mimesis.[35] Claude, however, is the son of Gervaise Macquart, a profligate, self-destructive alcoholic, who possesses and transmits the very hereditary taint which drives both Claude's obsessive drive to create, as well as his fundamental inability, in his view, to produce original work. To become an artistic success, therefore, would not merely enrich him such that he becomes a member of the bourgeoisie but in a more profound sense would index an overcoming of hereditary predestination and competitive selection. Claude also passes this taint on to his only son, a sickly, seemingly underdeveloped child, whose death from an unnamed illness occurs in the final stages of the novel. An image of the child's corpse forms the basis of Claude's final attempt at a successful Salon submission, entitled *Enfant mort*, a painting that seeks to represent, and thereby to overcome, the hereditary weakness that pursues both him and his son throughout the novel. The painting's morbid representation of the artist's dead son is met with public disgust and critical rejection. The novel concludes with the collapse of Claude's relationship with Christine, his one-time muse and the mother of his child, and with his eventual suicide. He is found hanged in front of his final, unfinished painting, the culmination of an intractable conflict between his desire to create original work and his incapacity to do so.

If Claude's failure in the struggle against nature is aesthetic, personal, and biological in nature, it can also be understood in social and political terms. His artistic practice and approach to realism (he is a proponent of the *en plein air* school) are proxies for the Impressionist movement, driven in a conscious way by contempt for academic aesthetics and, like Florent, bourgeois orthodoxy.[36] The theorisation of aesthetic practice in this novel also reveals larger, revolutionary ambition and speculates as to the capacity of his work to instigate profound change beyond the realm of the aesthetic. For Claude, a direct and unadorned form of realism and the representation of everyday subject matter contains revolutionary potential: 'Le jour venait où une seule carotte originale serait grosse d'une révolution' [The day was not far off when one solitary carrot might be pregnant with revolution!] (*L'Œuvre*: 43; *Masterpiece*: 41). In response, his interlocutor, Sandoz, offers an outline of a new kind of scientifically-motivated programme of painting. In an ironic reversal of the way genitality seems to stand in for biology in Lanoux's account of Zola's Darwin, Sandoz outlines his conception of a grand sequence of paintings, the title of which calls to mind Antoine Courbet's famous genital painting, *L'Origine du monde*. This

35 Quoted in William J. Berg and Laurey Kramer Martin, *Émile Zola Revisited* (Twayne, 1992), p. 196.

36 M. Douglas Kimball, 'Émile Zola and French Impressionism', *Bulletin of the Rocky Mountain Modern Language Association*, 23.2 (1969), pp. 51–57 (p. 51), doi:10.2307/1346694; William J. Berg, *The Visual Novel: Émile Zola and the Art of His Times* (Pennsylvania State University Press, 1992).

grand project, Sandoz proclaims, is intended to represent human life in all its evolutionary and cosmological dynamism:

> D'abord, épris de besognes géantes, il avait eu le projet d'une genèse de l'univers, en trois phases: la création, rétablie d'après la science; l'histoire de l'humanité, arrivant à son heure jouer son rôle, dans la chaîne des êtres; l'avenir, les êtres se succédant toujours, achevant de créer le monde, par le travail sans fin de la vie. (*L'Œuvre*: 44–45)
>
> [He had started by toying with the idea of a gigantic undertaking and had projected an 'Origins of the Universe' in three phases: the creation, established according to scientific research; the story of how the human race came to play its part in the sequence of living beings; the future, in which beings succeed beings, completing the creation of the world through the ceaseless activity of living matter.] (*Masterpiece*: 38)

Ultimately, Sandoz changes his mind, and instead seeks 'un cadre plus resserré, plus humain, où il ferait tenir pourtant sa vaste ambition' [a more limited, a more human setting for his ambitious plan] (*L'Œuvre*: 45; *Masterpiece*: 39) and in doing so seems to discard the quasi-Zolian socio-biological project that this calls to mind. This too is the scale of Claude's aesthetic vision for realism, at least in his *Enfant mort* and his final failed portrait of his lover, Christine, in front of the canvas of which she finds him hanged. His aim to paint 'things as they are' mirrors Sandoz's desire for totality, whose methodology differs only in its earlier claim to be scientific but not in its substantive aim for mimetic truthfulness. For Claude, it might be less the impossibility of such a project, but the hereditary taint, that determines his aesthetic and psychic breakdown, which suggests that in *L'Œuvre*, as in *Le Ventre de Paris*, revolutionary hope is necessarily punctured by the biological destiny that in turn reproduces the reigning bourgeois social, political, and aesthetic status quo — the fat devouring the thin.

These two novels mirror one another thematically and structurally. An individual marked by hereditary weakness, imbued with revolutionary energy, struggles against biological and social constraints to transform his conditions of existence, only for precisely that struggle to cause his downfall and death, an outcome seemingly predetermined by hereditary fate. In their efforts at transcending a realist image of socio-biological life, both protagonists, Claude and Florent, seem only to repeat the 'nausée de vertige' of hereditary pre-determination and the Darwinian war of nature. If *Germinal* is concerned with the potential for revolutionary change and meditates on its possibility in its narrative irresolution and dialogic reflection on social Darwinism, the social-Darwinian narrative logic in these other two works seem more definitively to crush such speculation in reaffirming social and biological homeostasis. Whether the attempt to transcend social and biological determination is through political action or artistic creation, failure seems to be the one possible outcome offered by these narratives. The question of social and political transformation and its biological impossibility in these works can be mapped onto Zola's theoretical effort at totalising scientific observation and representation and its necessary failure. *Le Roman expérimental* is an empirical and realist project in the service of a political project of reform: 'résoudre scientifiquement la question de savoir comment se comportent les hommes, dès qu'ils sont en société' [to solve scientifically the question of how men behave when they are in society].[37] Realist representation is understood by Zola, as well as Sandoz and Claude Lantier, as a catalyst for change, whether reform (as in the case of Zola) or revolution (for Zola's pathetic revolutionaries). By depicting the inevitable failure of such a project, *L'Œuvre* acknowledges the impossible standard of objectivity to which its own theoretical framework is believed to aspire (and for which Zola's work is sometimes dismissed). Here again, as in *Germinal*, the novel elaborates an awareness of its own pseudo-scientificity.

Reading this meta-fictional gesture in light of its social-Darwinian subtext, we might understand it as a reaffirmation of evolutionary determinism. Revolution, even reform, is impossible, for Sandoz and Claude, not only because humanity is irredeemably bound by a struggle for existence, but also because evolutionary change undermines the realist project of totalising mimesis. Yet, if *Germinal* can be read as a meditation on the political and historical implications of adopting a Darwinian theory of revolution, *L'Œuvre* can be interpreted as a narrative that charts the impact of a desire for evolutionary realism on the artist. This does not negate the narrative pessimism of the novel, but suggests we read Claude's failure and death in a more contingently negative fashion. Claude's ending, like Étienne's, is a symptom of different kind of fate, an elaboration of theoretical and practical failures of evolutionary aesthetics and politics respectively. Florent, in *Le Ventre de Paris*, by contrast, has no such intellectual pretensions; his plans for revolution are based on no explicit theoretical model or philosophy. Yet, he too can be understood as seeking to transcend his own conditions of biologicity, even if, unlike Étienne and Claude, the resources he employs to do so are not themselves informed by Darwin or biological thought. As such, in these works, the end of the human, as a telos, is not simply a question of transcending the biological (or effacing one's traces) but of the task of elaborating new artistic and intellectual relations with one's evolutionary predicament.

La fêlure: Death Instinct and Evolution

The productivity of the end: this is the theme of Gilles Deleuze's attempt in *The Logic of Sense* to read Zola's fiction of biological determinism in a radically optimistic way. Deleuze seeks to articulate a conception of hereditary transmission through Zola's work that is adequate to the indeterminacy of biological life, an indeterminacy which Deleuze also claims is the revolutionary core of Darwin's thesis in *On the Origin of Species*. In this reading, the bathetic negativity and violence of Zola's work (as characterised in these novels by massacres, suicide, starvation, and forced exile) are not merely symptoms of an evolutionary and hereditary pessimism that demands human transcendence. Rather, it is precisely catastrophe, violence, and struggle which reveal a force of transformation that goes beyond the immediate representational level of Zola's fiction and its vulgar determinism.

37 Zola, *Le Roman expérimental*, p. 20; *The Experimental Novel, and Other Essays*, trans. by Sherman, p. 21.

Deleuze's interpretation of the *Rougon-Macquart* project, which is founded on a reading of Zola's novel *La Bête humaine*, revolves around the distinction he makes in Zola's writing between two separate but interacting layers of narrative movement: that of the '*soma*' and of the '*germen*'.[38] In Deleuze's telling, the somatic layer contains the dramatic or historical tier of events in the plot: Claude's death and that of his son; Florent's exile; the massacre of the strikers and Souvarine's self-immolation; in other texts, murders, instances of adultery, alcoholic breakdowns, events that occur at the level of the human body, through which malignant forms of heredity are articulated. Beneath this lies an epic, germinal register of narrative that is once immanent to the somatic and goes beyond it. The 'adventure of men and their instincts', writes Deleuze, is played out in the 'small manoeuvre' of the tragic plot. Although linked, these narrative tiers are distinguished by their relation to human possibility: unlike the 'closed space' of hereditary destiny and the 'pagan character' of epic narrative elaborates an 'open space' of grand and supra-human narrative futurity, indeterminate and beyond human control.[39]

This dual narrative register, the germinal-epic and the historical-somatic, maps on to a two-fold conception of heredity. Somatic heredity transmits 'instincts' and well-defined pathologies such as alcoholism, perversion, and obsessionality from body to body. This could be applied to Étienne in *Germinal*, to Claude in *L'Œuvre*, and to Florent in *Le Ventre de Paris*, each of whose biological defects (transgressive political views, monomaniacal obsessionality, and a constitutional 'thinness') are apparently transmitted through familial lines of descent. The other form of heredity, which Deleuze calls 'la fêlure' [the crack], is germinal, a force of potential, which rests silently beneath the noise of the drives and the surface events which make up the plots of Zola's fiction. This 'crack', writes Deleuze, is the 'great internal Void' that traverses the *Rougon-Macquart* project and its characters, a force that transmits only itself, the possibility of transmission, and, thereby, the transmission of possibility. To grasp only the tragic, deterministic expression of biological heredity in Zola's work, argues Deleuze, is to conflate historical-somatic heredity with its vehicle: confusing the evolutionary tree of biological transmission with its drive for nothing other than its own continuation. This latter impulse is an empty topos, 'the crack', a void that connects the *Rougon-Macquart* genealogical network, the inherent possibility of collapse, which designates 'Death — the death Instinct'. Yet this is not the death instinct of neurotic repetition of the same; it is, for Deleuze, a topos of pure possibility, an 'imperceptible rift or hole' through which repetition shatters to become creative potential.[40]

38 Gilles Deleuze, *The Logic of Sense*, ed. by Mark Lester, trans. by Constantin V. Boundas and Charles Stivale (Columbia University Press, 1990), p. 320. This idea owes something to Mikhail Bakhtin's conception of the epic which Wai Chee Dimock connects to the scale of Darwinian narrative in 'The Survival of the Unfit', in *After Darwin: Literature, Theory, and Criticism in the Twenty-First Century*, ed. by Devin Griffiths and Deanna K. Kreisel (Cambridge University Press, 2023), pp. 121–37 (p. 124).

39 Deleuze, *The Logic of Sense*, p. 331.

40 *Fêlure* could also be translated as 'flaw', as in hereditary flaw or crack. I prefer to keep it as 'crack' (as Boundas does in his translation) to better communicate the difference between the surface effects of a flaw or taint in somatic heredity, and the subterranean issue of a crack or crack-up.

This rendering of the death instinct in Zola's fiction inverts some conceptions of the notionally Freudian schema of Eros and Thanatos, according to which the tendency towards life, reproduction, and creation is thought to counterpose aggression, self-destruction, and the compulsion to repeat.[41] Instead, for Deleuze, the death drive has a vital transformative potential while the instinct towards life, in so far as it is dedicated to survival (to use a term freighted with Darwinian significance), represents self-preservative tendency and repetition of the same. The instincts, temperaments, appetites, and obsessions that propel Zola's characters towards self-annihilation, Deleuze maintains, to the extent that they can be understood as failures to adapt in the war of nature, are also enactments of a more fundamental will to live. Although ostensibly self-destructive at the level of the individual, these tendencies are expressions of a type of life that seeks to elaborate itself 'in a historically determined existence in an unfavourable environment'.[42] In the case of Claude Lantier, for example, this means that his desire for what is an impossibly truthful form of mimetic art can be read not only as a self-destructive means of professional self-sabotage or the telos of a hereditary fate characterised by inherent struggle. Instead, we may see his ambition as an attempt at creating a transformative kind of life through art in response to a biologically or historically determined situation. Equally, Florent's engagement with political radicalism in Second Empire France need not be read as the unknowing instantiation of an unconscious desire for his own exile, even if for the reader it seems clear that his poorly concealed plans represent a wish to be found. Rather, according to this Deleuzean logic of the death drive, Florent's commitment, however delusional, is expressive of a desire for a life that he deems ethically acceptable, even if it is ultimately fatal for him: an affirmation of agency in response to the demands of rigid bourgeois organisation and respectability.

For all that Deleuze's re-elaboration of the death drive in relation to heredity suggests that these tragic protagonists represent radical forms of human agency, there seems little evidence of the kind of transformation or revolutionary change that might, in instrumental terms, justify their respective efforts. Claude's suicidal expression of aesthetic agency is doomed to repeat itself, seemingly never managing to achieve the transformation in aesthetic sensibility that could lead to or reflect a transformation in society. In the moments leading up to his suicide, he wonders aloud whether Eros, in the form of art, is within the grasp of human agency: 'Est-ce que donc impossible de créer? Nos mains n'ont-elles donc pas la puissance de créer des êtres?' [Is creation impossible? Are human hands powerless to make

Deleuze, *The Logic of Sense*, pp. 321–24.

41 Termed 'schizoanalysis' rather than 'psychoanalysis', Deleuze and Guattari's engagement with Freud is premised on a rejection of 'all reductive psychoanalytic and political analyses that remain caught within the sphere of totality and unity, in order to free the multiplicity of desire from the deadly neurotic and Oedipal yoke'. Part of this rejection consists in re-reading the death-drive as an affirmative constituent of the multiplicity of desire, which Deleuze prefigures in his reading of Zola in *The Logic of Sense*. See Mark Seem, 'Introduction', in Gilles Deleuze and Félix Guattari, *Anti-Oedipus: Capitalism and Schizophrenia*, trans. by Robert Hurley, Mark Seem and Helen R. Lane (University of Minnesota Press, 1983), pp. xv–xxiv.

42 Deleuze, *The Logic of Sense*, p. 322.

things come to life?] (*L'Œuvre*: 247; *Masterpiece*: 346). In suicide, Claude, it would appear, defaults to Thanatos. Similarly, Florent's plans for political insurrection seemingly produce no effect other than to solidify the bourgeois dedication to consumption and orthodoxy that his putative revolution seeks to challenge. La Belle Lisa, the family matriarch, reflects on the impact of Florent's arrival: 'c'est que nous n'avions plus nos habitudes [...]. Et la maison, non plus, n'a pas marché cette année. C'était comme un sort' [normal life had been disturbed [...] the shop hasn't been doing very well this year either. It was as if a spell had been put on it] (*Le Ventre*: 463–64; *Belly*: 271–72). Florent's deportation allows the Quenu family to return to equilibrium, to hope for the return of profitability, and for the nameless malaise that marked Florent's time in Les Halles to be replaced with normality. Nonetheless, argues Deleuze, the heredity of the crack, the vital death instinct, lies beneath all these events, imperceptibly binding each *Rougon-Macquart* novel into a single unity. For Deleuze, everything that occurs in these novels — the drama of familial conflict, murder, war, alcoholic dissolution — both covers over and points to this subterranean *epos* of the crack. It spreads itself throughout all the individual narratives and points towards a vast, supra-human narrative of life on a scale that exceeds and is innate to the lives and appetites of the human being. Keith Ansell-Pearson offers a sketch of the virtual topography of the crack and affirms its transformative potential. Alluding to the symbolic resonance of the steam train in Zola's work, he writes:

> On the tracks of this germinal train of life there is neither beginning nor end, neither a given genealogy nor a given teleology, but only the broken middles that allow for novel intersections, cross-connections, and unpredictable growths, constituting a cornucopia of good and bad. The crack enjoys a capacity for self-overcoming, making possible creative 'evolutions', in which the creation involved offers not a simple redemption but allows for the germinality of the most destructive inclinations and tendencies.[43]

The heredity of the crack transmits only transmission, communicates only the possibility of communication, and represents incalculable possibility as such: '[t]here remains [...] a future for heredity simply because what is inherited in any passing on is the future'.[44] As such, what we ordinarily take as the symptoms of repetition compulsion, self-sabotage, self-destruction, aggression, and so on, can be revised not as the expression of a repressed and self-defeating drive, but as a necessary condition for transformation, whether for good or for bad.

This radical revision of Freud's death drive in relation in Zola's novels has important implications for the way we read Zola's engagement with Darwin. In Deleuze's account, despite its notionally destructive character, the death drive cannot be linked merely to destruction or masochism. Nor, by extension, can it be easily associated with the version of Darwinian natural selection according to which

43 Keith Ansell-Pearson, *Germinal Life: The Difference and Repetition of Deleuze* (Routledge, 1999), p. 114.

44 Keith Ansell-Pearson, 'Spectropoiesis and Rhizomatics: Learning to Live with Death and Demons', in *Evil Spirits: Nihilism and the Fate of Modernity*, ed. by Gary Banham and Charlie Blake (Manchester University Press, 2000), pp. 124–46 (p. 132).

life is defined by a 'struggle to survive'.[45] The bathetic failures of the apparently doomed protagonists in Zola's novels, Claude Lantier, Étienne Lantier, and Florent Quenu, are not simply extrapolations of evolutionary mechanism or hereditary pre-destination but attempts to transcend these conditions through the enactment of political or aesthetic possibility. Indeed, it is the failure of these efforts that opens out onto the topos of pure potential that Deleuze calls 'la fêlure'. By contrast, the life instinct — the struggle to survive — is a conservative process by which everything remains static through repetition. We can see this, for example, in the demands that are made on Claude and Florent to comply with bourgeois normality, whether by producing commercially and critically palatable painting or maintaining a post as an inspector in Les Halles respectively — both roles that function to maintain a certain social, economic, and aesthetic status quo. In this light, it becomes possible to revise Zola's pessimism because, in its depiction of revolutionary failure, his narratives disclose this otherwise imperceptible topos of the death instinct:

> It is one and the same movement — the movement of the epic — that the basest instincts are reflected in the terrible death instinct, but also that the death instinct is reflected inside an open space [the crack], perhaps even against itself.[46]

Thus, interpretations of what Deleuze calls the 'socialist optimism' of Étienne in *Germinal* are misguided because they are wedded to a somatic conception of heredity (progress) that can only repeat the same. A more radically optimistic reading of Zola, Deleuze suggests, consists in recognising that the 'proletariat already makes its way through the crack', that life as such is always 'endowed with a future'. This conception of a movement of transformation that is always already underway, even at the expense of human life, is opposed to an illusion of redemptive Darwinian progress founded on a fantasy of a transcendent human subject for whom death can only be conceived of as a tragic failure.[47]

So Zola's work can be seen as reimagining Darwin's theory of natural selection and its political, social, and ethical ramifications. The two levels of narrative at work in Zola's fiction split apart the Darwinian model of evolution by natural selection, shifting the locus of transformation away from the violent antagonism of competition and extinction. The question of whether natural selection is the dominant mode of evolutionary relation or one of a number of dynamics emerges here, and will re-emerge again in the next chapter on Hardy's treatment of Darwinian sexual selection. For Elizabeth Grosz (and as I have suggested, for Freud), the agonistic field of relations circumscribed by natural selection is supplemented by another dynamic in the form of incalculable sexual desire and its cultural formations (which we might more readily associate with Eros). Here, however, for Deleuze, it

45 Julia Reshe offers a useful and more explicitly psychoanalytic comparison of Darwinian evolution and the Freudian death-drive that comes ostensibly to the opposite conclusion, positing a form of 'depressive realism' as against Deleuze's revelatory fiction of the crack: Julie Reshe, 'The Death Drive of Evolution (from the Perspective of Depressive Realism)', *Stasis*, 11.1 (2021), doi:10.33280/2310–3817–21–11–1–156–180.

46 Deleuze, *The Logic of Sense*, p. 332.

47 Ibid., p. 324.

is precisely the void of Zola's Thanatos — not the presence of a creative Eros — that guarantees transformation, which works silently beneath the noisy events of mere survival. Not only, therefore, does this tear open the Darwinian model of evolution, it also reorients our reading of the dynamics of natural selection, the representation of which violent mechanisms can be read (not unproblematically, as we shall see) as a way of sensing the *epos* of incalculable change.

These levels of transformation are evident at the conclusion of *Germinal* and its figuration of the subterranean as the topos of both mass death and futurity. As Étienne leaves Montsou, soon after questioning the Darwinian foundations of his own revolutionary theoretical disposition, he intuits that the morbid failure of the political movement represents a precondition for vital biological and political regeneration. He imagines the sound of coal picks beneath him forming a chorus of vital revolutionary possibility, the tapping of his defeated comrades providing the rhythm for a germinal renewal of life after the horrors of political failure and death:

> Aux rayons enflamés de l'astre, par cette matinée de jeunesse, c'était de cette rumeur que la campagne était grosse. Des hommes poussaient, une armée noire, vengeresse, qui germait lentement dans les sillons, grandissant pour les récoltes du siècle futur, et dont la germination allait faire bientôt éclater la terre. (*Germinal*: 553)
>
> [Beneath the blazing rays of sun, in that morning of new growth, the countryside rang with its song, as its belly swelled with a black and avenging army of men, germinating slowly in its furrows, growing upwards in readiness for harvests to come, until one day, soon, their ripening would burst open the earth itself.] (*Germinal*: 524)

The swarms of workers beneath his feet grow in readiness, not for a pre-given ascension to victory, but instead, as Zola has it in the novel's final line, for 'les récoltes du siècle futur'. The term *récolte* rather than *rassemblement*, to which it is related via the Latin *recolligere*, underlines the way in which this final passage frames revolutionary transformation as both organic and subterranean. The novel ends, therefore, with morbid failure and slaughter, notionally determined by the mechanism of Darwinian natural selection and heredity, opening up an indeterminate, organic sense of vital possibility. The collapse, the crack-up, of the strike is not an end to the possibility of transformation but the moment at which we can glimpse, in its dissolution into a subterranean void of pure difference, the possibility of a change whose cause and telos exceeds that of the human.

This dynamic is staged too in *L'Œuvre* in another concluding scene, another notional ending. Claude's funeral ostensibly represents the final punctuation mark in the novel's biologically deterministic narrative, the inevitably tragic life of an artist, doomed by his heredity to sabotage his struggle for survival. In attendance is Bongrand, one of Claude's fellow artists, and Sandoz, the fictional proxy for Zola. The former begins a eulogy not for Claude but for progress in the nineteenth century which, not only calls to mind Ortega's lament for the nineteenth century, but also the triumph of a destructive death instinct: 'Ah! Oui, l'air de l'époque est Mauvais, cette fin de siècle encombrée de démolitions, aux monuments éventrés, aux terrains retournés cent fois, qui tous exhalent une puanteur du mort!' [We're living in a bad season, in a vitiated atmosphere, with the century come to an end and everything in process of demolition; buildings torn down wholesale; every plot of land being dug and redug and every mortal thing stinking of death] (*L'Œuvre*: 255; *Masterpiece*: 359). For Sandoz, the revolutionary epistemic and aesthetic promise of science has failed to overcome this widespread sense of decay. He takes up Bongrand's theme and continues:

> C'est une faillite du siècle, le pessimisme tord les entrailles, le mysticisme embrume les cervelles; car nous avons eu beau chasser les fantômes sous les grands coups de lumière de l'analyse, le surnaturel a repris les hostilités, l'esprit des légendes se révolte et veut nous reconquérir, dans cette halte de fatigue et d'angoisse. (*L'Œuvre*: 255)
>
> [The century has been a failure. Hearts are tortured with pessimism and brains clouded with mysticism for, try as we may to put imagination to flight with the cold light of science, we have the supernatural once more in arms against us and the whole world of legend in revolt, bent on enslaving us again in our moment of fatigue and uncertainty.] (*Masterpiece*: 359)

These eulogies for Claude and for the promise of scientific progress can be read as a statement of Zola's deep pessimism regarding the entropic dynamic of biological and social life and the capacity of art to overcome it. As Claude is lowered into the grave, the funeral service is drowned out by 'une grosse locomotive [...] en soufflant' [a huge, puffing locomotive] (256; 361), an emblem of industrial progress used by Zola throughout the *Rougon-Macquart* as a symbol of the interconnectivity and inexorable movement inherent in heredity. Heredity and progress drown out even attempts to reflect on its inevitability.

However, for Deleuze, the train as a symbol in Zola's works represents germinal and not somatic heredity — 'that which rushes by, a mobile spectacle linking the whole earth and men of every origin' — not biological predetermination but precisely its opposite: the very 'image of death'. That is, in Deleuze's inverted sense, 'the pure Death instinct', a symbol of the epic heredity of the crack that joins together Zola's novels and represents that which precedes and exceeds narratives of death, crime, and adultery.[48] Claude's death then, as a synecdoche for a wider entropy, represents both the end of life and points towards that which creates it: pure, immanent possibility. For Sandoz, too, 'Nous ne sommes pas une fin, mais une transition, un commencement d'autre chose' [We are not an end; we are at a transition, the beginning only of something new] (255; 359). Here, echoing the conclusion of *Germinal*, *L'Œuvre* narrates at length the hereditary origins of failure and death, only to posit at the conclusion of the novel that the end represents nothing other than a transitional point in evolution.

48 Ibid., pp. 329–30.

Putrid Art: 'the Proteus of generation'

Deleuze's concept of a somatic and germinal heredity mirrors that of the nineteenth-century doctor and theorist of heredity, Prosper Lucas, whose work was instrumental in Zola's conception of heredity. Like Deleuze, Lucas divided heredity in two: the law of imitation, or sameness, which powers inheritance, and the law of *innéité*, or difference, often translated as 'variation' or 'mutation'. The first form of heredity ensures the transmission of characteristics (or 'instincts'), such as mental diseases; the latter, theorising in the absence of genetic conceptions of mutation and inheritance, represents a more diffuse form of transmission that brings forth the possibility of variation and change. Echoing Deleuze, Lucas writes of this latter form, *innéité*, that it is the 'veritable Proteus of generation [...] the very incarnation of the diverse in life' which 'progress[es] by transforming all the elements, all the attributes, all the modes of life on which its action has a bearing'.[49]

Darwin too acknowledged Lucas's influence in *On the Origin of Species* but folded together, under the rubric of natural selection, the processes of hereditary transmission through selection and the transmission of differences.[50] Variation through the production of difference, Darwin speculates, is a response to an organism's physical conditions of life, the somatic. He recognised in *On the Origin of Species* the physiological plasticity of the organism, but attributed its variations to the effects of selection and adaptation:

> But the much greater variability, as well as the greater frequency of monstrosities, under domestication or cultivation, than under nature, leads me to believe that deviations of structure are in some way due to the nature of the conditions of life, to which the parents and their more remote ancestors have been exposed during several generations. (*Origin*: 131)

By contrast, for Lucas and Deleuze, the corporeal world produces only sameness; variation is not located in the world of physical differences or adaptations, but in the Protean ontological condition of difference that precedes the physical materialisation of differences (innateness vs innate characteristics). Zola's work, then, could be read as combining these two models of heredity and evolution, Lucas's theory and the Darwinian mechanism, by conjoining the social and political dynamics of natural selection and the Protean condition of difference in its depictions of revolutionary failures, broken ends, and crack-ups. This has the effect of dividing Darwin's conception of heredity through natural selection, as outlined in *On the Origin of Species*, to reinstall the primacy of difference (as an ontological condition that pre-exists concrete differences) and to relocate natural selection as part of a multiplicity

49 Translation by David Baguley, taken from his 'Darwin, Zola and Dr Prosper Lucas's "Treatise on Natural Heredity"', p. 428, originally found in Prosper Lucas, *Traité philosophique et physiologique de l'hérédité naturelle* (Paris: J. B. Baillière, 1847), p. 649.

50 'Any variation which is not inherited is unimportant for us. But the number and diversity of inheritable deviations of structure, both those of slight and those of considerable physiological importance, is endless. Dr Prosper Lucas's treatise, in two large volumes, is the fullest and the best on this subject' (*Origin*: 13). For an outline of Lucas's influence on Darwin see Ricardo Noguera Solano, 'Darwin and Inheritance: The Influence of Prosper Lucas', *Journal of the History of Biology*, 42.4 (2009), pp. 685–714, doi:10.1007/s10739-008-9175-7.

of evolutionary processes. For Deleuze, Zola's tragic failures disclose a space beneath and in excess of the vicissitudes of the social-Darwinian struggle for life, recasting death, not as extinction, but as futurity's condition of possibility.

In doing so, both Zola, in his *Le Roman expérimental*, and Deleuze, in his reading of heredity in Zola in *The Logic of Sense*, ascribe particular kinds of agency to literary art. For Zola, the capacity of art (and science) to represent reality makes it possible to recruit literature into the service of a positivist scientific programme of observation, representation, and amelioration. The way Zola's fiction seems to question that theoretical stance, offering little in the way of ameliorative optimism at the level of narrative action while thematising and questioning its own scientificity, outlines a vision of human agency which is limited to critical reflection. Deleuze's interpretation, by contrast, suggests that Zola's work has a very different, but powerful, function: to unveil the absolute anti-determinism of biological life in its presentation of heredity. In this light, the positive project of Zolian Naturalism is neither prognostic (in a deterministic sense) nor only diagnostic (medically or existentially), but revelatory.

To instantiate this way of reading Zola, Deleuze suggests we focus on the most violent and 'putrid' aspects of fiction through which, according to his articulation of *la fêlure*, it is possible to intuit a sense of the infinite *epos* of the death drive. Contrary to conceptions of Zola's novels as too *lisible*, for Deleuze it is in his works' exaggerated and descriptive hyperbole, used in the depiction of the 'history of the instincts against the background of death', that the epic character of the heredity of the crack is revealed. A writer of epic literature, Deleuze says, 'has no *logos*, but only an *epos*', in which there is no appropriate level of realism.[51] In other words, Deleuze suggests that it is precisely in the most violent, most morbid parts of Zola's fiction, where it is at its most dramatic or hyperbolic (as opposed to objective), in which a somatic connection with the germinal epic can be grasped. To grasp something of the morbid 'real' of the post-Darwinian life, it is implied, calls for exaggeration. This revises the ideological critique of realism, according to which a confrontation with the notional 'realities' of biological and social life does not transform but functions to naturalise those conditions.[52] For after Deleuze, we might also add that to falter in reflecting the abject cruelty and horror of biological life, to skim over the 'basest instincts', indeed to fail to exaggerate these, is merely an attempt to protect ourselves from the prospect of the bodily and affective consequences of our own mortality. It is, to return to Freud's gloss on Darwin, a way of avoiding the mercilessly inhuman nature of the post-Darwinian conception of nature (and, in Zola, society). As such, averting our gaze from the inhuman also permits us the fantasy of representational control over the symbolic death of the human at the hands of scientific revolutions (such as that of Darwin).

If in *Germinal* and *L'Œuvre* this revelatory putridity is represented by political

51 Deleuze, *The Logic of Sense*, p. 332.

52 'The classic realist text installs itself in the space between fact and illusion through the presentation of a simulated reality which is plausible but not real. In this lies its power as myth': Catherine Belsey, *Critical Practice*, 2nd edn (Routledge, 2003), p. 96.

violence, mass death, and suicide, in *Le Ventre de Paris* the apparently banal site of the bourgeois kitchen is the locus of an otherwise hidden violent abjection. This is emblematised by the story, recounted by Étienne in his family's *charcuterie*, of his escape from exile on Devil's Island. '"Dis, cousin, raconte-moi l'histoire du monsieur qui a été mangé par les bêtes"' ['Tell me the story of the man who was eaten alive!'], enjoins his niece, as in the kitchen the preparation of a *boudin noir* is described by the narrator in precise Naturalistic detail. The narrator makes explicit the connection between Florent's tale and the context in which it is told: 'Sans doute, dans cette tête de gamine, l'idée du sang des cochons' [No doubt the mention of blood flowing from pigs] instigates the niece's interest in this tale, tying together Florent and the entrails of a pig through the niece's association of the two. The tale begins with a description of Florent's experience of imprisonment on Devil's Island. There, the prisoners live in a Hobbesian state of nature and Malthusian scarcity, fighting a Darwinian battle for survival against mosquitoes, whose bites are fatal to numerous men, as they compete with maggots infesting what little food they possess. Reduced to the condition of an unbearable human-animality, the prisoners' struggle with the violence of the prison-camp guards is not merely a struggle for survival but to be acknowledged as human. Eventually Florent details how, having left one of his companions he and his fellow prisoners return to find him dead:

> Quand ils revinrent à l'écueil, ils virent leur compagnon étendu sur le dos, les pieds et les mains dévorés, la face rongée, le ventre plein d'un grouillement de crabes qui agitaient la peau des flancs, comme si un râle furieux eût traversé ce cadavre à moitié mangé et frais encore. (*Le Ventre*: 311–13)
>
> [When they got back to the rock they saw their companion lying on his back, his hands and feet eaten away, his face gnawed, and his stomach full of crabs, crawling about, making his sides shake, as if the half-eaten corpse, still fresh, was in the throes of a terrible death agony.] (*Belly*: 83)

This image to which Florent's niece alludes is emblematic of Florent's life in exile and the condition of human-animality to which he is subjected. It can also be read as an implicit commentary on the novel as a whole: here is the true condition of the 'belly' of Paris, a body consumed by the inhuman, where humanity is locked in a constant battle with a nature that is beyond domination. The description of the corpse is suggestive: it is consumed from within and without, as if it is the condition of animality that traverses humanity which leads to his demise. What is more, this internal-externality is also what seems to enliven the corpse, if only momentarily: the human becomes a kind of marionette, a performance conducted by the creatures that destroy him. If humanity is, as I have suggested, a performative entity, here life itself is a gruesome charade as its very trembling conceals a deeper morbidity from within. Yet, as Deleuze suggests, we may also view this encounter with human-animality as a way of intuiting the topos of the transformative death drive. Perhaps its attendance to the non-human, the life force of the crab, challenges a reflexive anthropomorphism, by underlining not so much the brutality of natural selection, but the capacity for non-human life to metabolise what for us as humans reads as tragedy. After all, it is not the crabs that destroy this unfortunate prisoner, but the authoritarian revanchism of Second Empire France.

Moreover, the narrative-framing device of the preparation of the *boudin noir*, the process of which interrupts Florent's own narration, implies that savagery is less a feature of the non-human or the inhuman, but is constitutive of the bourgeois comfort enjoyed by the Quenu family. As Florent recounts this tale of biological warfare, his family and a kitchen servant congregate over vats of bubbling pig's blood, taking pleasure in the gory but sensuous processes by which the *boudin* is prepared: 'La bouillie coulait, toute noire et toute fumante, gonflant peu à peu le boyau, qui retombait ventru, avec des courbes molles' [The mixture, black and steaming, flowed through the funnel, gradually swelling the skin, which fell back in a fat, soft curve] (*Le Ventre*: 314; *Belly*: 87). While Florent details how the political exiles of the French state become the victims of nature's war, through starvation and deprivation, the family delight in the blood of another animal, the pig, an ironic symbol of their own capitalist greed, whose steaming fat seems to infuse the very atmosphere in which they reside. Both the pig and Florent's companion (and, as we know, Florent himself) are presented therefore as victims of the common underlying antagonism of the fat and the thin, the strong and the weak.

However, the subject's position within this dyad depends not on a pre-destined flaw or hereditary mark, but on their relation to the coordinates of the struggle. For Florent, if only briefly, the attraction of becoming a member of the well-adapted bourgeoisie overcomes what at this point in the novel is still an internal revolt against middle-class orthodoxy and injustice: 'Il glissait à la lâcheté heureuse de cette digestion continue du milieu gras oû il vivait depuis quinze jours. C'était, à fleur du peau, mille chatouillements de graisse naissante, un lent envahissement de l'être entier, une douceur molle et boutiquiére' [He was falling into the state of blissful complacency created by the continuous focus on food and well-being that defined the world in which he had been living for the last two weeks. He felt, as it were, the titillation of fat forming all over his body. His whole being was slowly being invaded by the languid contentment of a shopkeeper] (*Le Ventre*: 316; *Belly*: 89). Rather than be devoured from within by animality, Florent instead allows himself to be invaded by the pleasure of bourgeois comfort and the possibility of satiation. This is a form of survival; however, as Deleuze makes clear, this is not a form of creative life instinct but its opposite. If the Oedipal and bourgeois comforts of the Quenus' commercial kitchen are to be understood as expressions of Eros — of life, survival, preservation — it is a conventional type of life-instinct centred on the reproduction of the nuclear family and the production of surplus value. The narrative frame, the realist preoccupation with pig's blood, underlines the way this bourgeois life instinct, dedicated to the maintenance of a social and political status quo, is contingent on the very kind of violent slaughter and symbolic consumption that defines the Darwinian war of nature — necessarily disavowed to uphold a sense of middle-class respectability. On the contrary, Florent's tale, in its explicit acknowledgement of the somatic consequences of political repression as well as the violence of human-animal life, seems to instantiate precisely this dialectic of concealment/revelation in relation to the epic, germinal tier of transformation. The 'man who is eaten alive' appears to us both as a repugnant exemplar of 'ineradicable animal nature' and, in so doing, indexes (perhaps against our will) a more profound, supra-human form of life and vital transformation, contingent on abjection and failure. The presence of death both on Devil's Island and in the Quenus' kitchen, moreover, is linked to this: there is consolation in the notion that if the thin or weak are fated to lose in the struggle for survival, so too — despite a perverse disavowal of this fact — are the fat and the strong.

The germinal heredity of the crack as theorised by Deleuze, then, allows us to read Zola's putrid literature in affirmative but not redemptive terms while refusing a morally humanistic understanding of life. For some readers of Deleuze, this is too great a demand: Deleuze's reading of Darwin, it has been argued, in ratifying the ontological primacy of 'difference in itself', ignores the brutality of the processes of selection through which difference is distributed, whether those are social or biological forms of change.[53] Yet, Deleuze's reading of Zola is not an invitation to ignore or disavow violence in favour of vital difference, but to attend to the way in which the two are connected. Although Deleuze's wider intellectual project is aimed at working beyond representation, the double narrative tiers of Zola's fiction nevertheless suggests a kind of realism that stages a dynamic organic totality. Zola's writing intimates that nature is constituted by both death and growth, by both negation and abundance, by a conservative heredity and the potential of a violent collapse into creative transformation. His pathetic revolutionaries are avatars of this dynamic, whose suffering or failure is never resolved by the narrative, is in fact aggressively exaggerated, thus forcing us to relinquish the false hope of redemption in favour of a revolutionary capacity to act as if we were undetermined human subjects, even if in the face of destruction.

The psychoanalytic theorist and critic Julia Kristeva also identifies in Zola's fiction, contrary to the contemporary banalisation of violence, a confrontation with the brutality of human experience.[54] Reading Zola, Kristeva suggests in a brief but suggestive note, 'Aimer la vérité cruelle et disgracieuse', that she finds a mirror of the social-Darwinian war of nature and the symptoms of pre-determined hereditary taint: 'la violence crue et disgracieuse de la sexualité brutale, de la pathologie nerveuse, de la détresse, banale et pour cela même cruelle, des pauvres' [the raw and ugly violence of brutal sexuality, of nervous pathology, of the distress, banal and yet cruel, of poverty'].[55] Its function, Kristeva says, is to transcend the mimetic or indeed the aesthetic: 'En cela, cette œuvre échappe à l'esthétique et à ses critères, et nous installe dans cette région obscure de l'expérience humaine où l'imaginaire a partie liée à la clinique et à la métaphysique' [This work goes beyond the aesthetic and its criteria and installs in us a space of obscurity where the imagination is linked to the clinical and the metaphysical].[56] As such, in this assessment of Zola, Kristeva

53 Hallward, *Out of This World*, pp. 162–63; Caygill, 'The Topology of Selection'.
54 Julia Kristeva, 'Aimer la vérité cruelle et disgracieuse', *Cahiers Naturalistes*, 68 (1994), pp. 7–9.
55 Ibid., p. 8.
56 Ibid., p. 9.

shares Deleuze's focus on the revelatory impact of what is putrid in his fiction. For Kristeva, this approach is part of a wider theorisation of 'the abject' as a form of the sublime (or 'sublimation'), according to which ('great') literature's encounter with violence, death, or the 'unclean' has a cathartic therapeutic function — leaving a wound that cuts through the fantasy of egoic sovereignty to destabilise or vitiate the subject. If for Deleuze, Zolian fiction reveals the force of the supra-human, we might imagine that, for Kristeva, it works at the level of the human to penetrate the disavowal that otherwise characterises our relation to the decentring of the human. The answer, as we shall see, lies in whether the resistance activated by the abject in Zola's socio-biological realism indexes a slashing through the human subject's constitutive defensive posture, or functions in the opposite way, to shore up the human's psychic and discursive boundaries.

For Deleuze, the key to reading Zola's 'putrid' literature lies in an inversion of the Freudian psychoanalytic schema of life and death drive. Kristeva's notion of 'abjection' posits a form of corporeal and psychic repudiation which also seeks to go beyond standard Freudian accounts of pleasure and displeasure. Extreme disgust in Kristeva is neither limited to the neurotic symptom or psychotic deviance but is universal to the formation of normal subjectivity: a pre-intellectual, originary response to the trauma of parturition.[57] The infant, Kristeva posits, emerges from a space that precedes the subject and the object, being and non-being, and responds with horror to the prospect of this abyssal topos whose only attribute is the negation of the subject and its instantiation in language. As such, subjectivation is fundamentally a product not of desire (or creation) but of negation and the void, driven by terror and disgust at what is abject. 'The abject' for Kristeva is therefore not an object or symbol, but an intolerable and exorbitant topos whose only quality is that of being 'opposed to I'. The abject, Kristeva writes:

> Is [n]ot me. Not that. But not nothing, either. A 'something' that I do not recognize as a thing. A weight of meaninglessness, about which there is nothing insignificant, and which crushes me. On the edge of non-existence and hallucination, of a reality that, if I acknowledge it, annihilates me.[58]

Deleuze's reading of Zola articulates a conceptual narrative space of the death drive that lies beneath the corporeal and the vicissitudes of somatic heredity. By contrast, Kristeva's abyssal space of non-signifying significance is embodied, specifically by the object of the corpse, and by its wounds and emissions. Unlike medicalised or scientific representations of death, Kristeva argues, the direct encounter with a cadaver, a laceration, blood, pus, or the acrid smell of decay confronts us directly with what we as subjects 'permanently thrust aside in order to live'. This pure experience of the abject — of that which lies beyond the subject 'I' and its location in a binary relation to an object — is itself resistant to being objectified such that we might cast ourselves as a stable subject. Instead, in its presence we are threatened

57 Winfried Menninghaus, *Disgust: The Theory and History of a Strong Sensation*, trans. by Howard Eiland and Joel Golb (State University of New York Press, 2003), p. 367.

58 Julia Kristeva, *Powers of Horror: An Essay on Abjection*, trans. by Leon S. Roudiez (Columbia University Press, 1982), p. 2.

and harried by the company of non-existence and brought to 'the border of [our] condition as a living being'.[59] The consequences of Darwinian natural selection, in Freud's telling, has a similar effect, confronting the subject with the 'fact' of its animality. However, if in a Freudian account this produces resistance in a recognisable form of outright rejection (or a concealed rejection of its central radical anti-human postulate hidden in its apparent embrace), for Kristeva this encounter with the abject brings us to the very limit of our existence as human subjects.

In this light Claude's suicide in Zola's *L'Œuvre may* take on a new complexion. The suicide is caused, we are given to assume, by a nameless hereditary lesion, existing beyond signification, which both inspires his creativity and makes him incapable of accepting its limits. Deleuze, as I have shown, might cast this tragic narrative of biological fate in affirmative terms: Claude, faced with the pre-determined social and biological fate of somatic heredity, dedicates his life to a transgressive obsession with the creation of truthful art, an obsession which falters at the edge of its possibility until it turns back to death, and the germinal heredity of the crack. Kristeva's theory of disgust also reworks the Freudian notion of the death drive, suggesting that Claude's misery and suicide represent a logical horror at the groundlessness of his own art and his own biological constitution. Another way of articulating this is that it is both the impossibility of representational transparency and thereby the impossibility, as it appears to him, of transcending his own biological conditions that simultaneously attract Claude and repulse him. Appalled by his own inability to produce work that is adequate to his revolutionary ambitions, Claude returns repeatedly to this impasse, neurotically reconfronting himself with the impossibility of his own artistic project and his apparently inevitable biological fate. In one example, Claude becomes obsessed with creating a panoramic landscape painting of the Île de la Cité, the perfection of which consumes his life, and which, despite Herculean efforts, remains unfinished. He refers to this project as a battle with 'le Réel', the narrator alerting us to the 'impossible task of putting all nature on one canvas' which defeats Claude at every attempt. In response to that (and to the failure of his portrait of Christine) Claude's final act is to achieve a final form of self-determination in himself becoming abject.

Claude's attraction to death is crystallised in the work he paints in the moments after discovering his child's corpse. Finding his child deceased, killed by an unnamed hereditary condition, Claude is immediately struck by a nameless fascination with this corpse and a desire to paint:

> Et, quand il passait devant le petit cadavre, il ne pouvait s'empêcher de lui jeter un regard. Les yeux fixes, grands ouverts, semblaient exercer sur lui une puissance. D'abord, il résista, l'idée confuse se précisait, finissait par être une obsession. Il céda enfin, alla prendre une petite toile, commença une étude de l'enfant mort. (*L'Œuvre*: 193).

> [Every time he passed the child's dead body he felt obliged to look at it, as if the glassy, staring eyes were exercising some kind of power over him. He tried to resist it at first, but the attraction grew stronger and stronger to the point of

59 Ibid., p. 3.

> obsession, until at last he gave way, fetched out a small canvas and set to work on a study of the dead child.] (*Masterpiece*: 262–63)

Claude is bewitched by this the body of his dead son, this non-object, which draws him towards it, even as, frozen in death, it horrifies him with its deformed features: 'Ce dessin exagéré de la tête, ce ton de cire des chairs' [The exaggerated shape of the head, the waxlike texture of the skin]. What Claude ultimately perceives in the corpse of his child, in addition to an expression of hereditary weakness, is the abyssal emptiness of the abject, 'ces yeux pareils à des trous sur la vide' [eyes like holes wide open on the void] (193; 263). In Kristeva's terms, Claude sublimates this close braiding of contrary affects into art, which represents one of the few works that satisfies his aesthetic aspirations. The work, after much political manoeuvring, is accepted as a 'charity' submission to be exhibited at the Salon where its macabre subject matter stands in stark contrast to the works that surround it. Zola's ekphrastic prose describes the picture with as little restraint as Claude exercises in his own rendering of his dead son on to canvas:

> Le misérable petit cadavre, qui n'était plus, à cette distance, qu'un confusion de chairs, la carcasse échouée de quelque bête informe! Était-ce un crane, était-ce un ventre, cette tête phénoménale, enflée et blanchie? Et ces pauvres mains tordues sur les linges, comme des pattes rétractées d'oiseau tué par le froid! (*L'Œuvre*: 211)
>
> [Hanging where it did it was just a confused mass, like the carcass of some shapeless creature cast up by the tide, while the abnormally large head might have been any white, swollen object, a skull or even a bloated belly, and the wizened hands on the shroud looked like the curled-up claws of a bird that has died of cold.] (*Masterpiece*: 290)

This dark and surreal work is predictably unpopular with the crowds that visit the Salon; *Enfant mort* attracts moralising derision from judges and histrionic laughter among the crowds of public visitors. In Kristeva's conception of abject art, in its failure the painting is a type of success. For it represents this death in a way that is similar to Florent's depiction of the 'man who was eaten alive': not merely producing an affect of disgust but troubling in its abjection. The man and the child are both human and animal; turgidly alive and morbidly empty; the wreckages of a supra-human biological force. *Enfant mort* is as an image representationally opaque, shapeless, unformed, devoid of objectal boundaries. In this way, the work may function for the exhibition-goers in the same manner as it does for Claude: as a means by which one can confront the abject, to encounter that which stands in opposition to 'I', and become wounded, at the threshold of one's own being.

The repulsion that its viewers feel at the Salon at which the painting is exhibited appears to confirm this reading. As Kristeva might predict, the most radically wounding art is also that which 'the professional establishment, along with the cynicism of the times and of institutions, will soon manage to close up'.[60] Here, Zola's work in its presentation of a quasi-Darwinian abject depicts resistance not

60 Ibid., p. 27.

for aesthetic but for more profound psychic reasons of aversion. For Claude's *Enfant mort*, as for Darwin's natural selection, its reception indexes a deeper destabilisation of the distinctions of subject-object, human and animal, which constitute the concept of humanity as a biologically sovereign entity. And yet, while we might like to assume that a certain psychic dislocation has been experienced by the disdainful Salon-goers, this repudiation may in fact, again, reproduce an aesthetic and social status quo. If *Enfant mort* is an example of the art of abjection in theory, its ridicule in the context of the novel functions to ensure art of this type remains marginal and excluded from established institutional success. Similarly, if the 'man who was eaten alive' in Florent's story functions to wound its listeners, it does so only in the abstract: it is this tale that convinces the Quenu family of Florent's essential deviance. A confrontation with the Darwinian *struggleforlife*, for Lisa and her family and the bourgeois crowds of the Salon, merely strengthens their belief in the immortal health and vigour of the bourgeoisie and distrust of the biologically and socially vulnerable.

This ambivalence is anticipated in Kristeva's analysis, which is dedicated also to revealing how fascination and yearning are drawn by horror at the same time as this horror evokes our repugnance. Disgust, nausea, the symptoms of bodily and emotional repudiation: these powerful affects of aversion permit us to consolidate ourselves in opposition to the boundlessness of death. In addition to death, the loathing of food, turning away from and repudiating what is supposed to nourish us, is also a fundamental embodiment of the process of abjection, which in Kristeva's theory is linked closely to infantile subject formation. For example, a child confronted with exorbitant parental desire in the form of food refuses to assimilate this dependency, and in gagging, vomiting, or refusing nourishment, becomes other. But abjection is not represented by the act of refusal itself, since the horror of food or a corpse is not a response to an object per se. Rather, in abjection, Kristeva writes, 'I expel *myself*, I spit *myself* out, I abject *myself* within the same motion through which "I" claim to establish *myself*'.[61] So we are attracted by the abject in the same movement as we are disgusted by it because it offers us the particular kind of pleasure of abjection accompanied by disgust. Here, Claude's obsession with his son's corpse is instructive, as is Florent's niece's prurient interest in her uncle's suffering. For Kristeva, human subjectivity is perversely ambiguous: constituted both by desire for wholeness and in response to horror, subsisting on the intertwining of these two supposedly contrary affects. If these figures seem drawn to death or violence, it may be in part because such an encounter promises to consolidate rather than wound the ego. This reading might also trouble the Deleuzean narrative of the crack in Zola's fiction: rather than reveal a subterranean space of transformative possibility, its depiction and formalisation of violent antagonism and its tragic consequences provokes repulsion and attraction precisely because it allows us to reconstitute ourselves as subjects in response.

This chapter's analysis of biological negativity (the putrid and the abject) has centred on two images of death: the corpse of a political exile, related to us via

61 Ibid., p. 3.

Florent's story, and that of a small child, rendered in Claude's painting, each one a miniature artwork within Zola's wider narratives. These works, I have argued, may be read synecdochally, as metonymic representatives of the social-Darwinian dynamics that operate in the wider narrative logic of the novels. Yet, according to Kristeva's elaboration of the abject, we cannot simply read these nested artworks or their narrative containers as indexing a sublime of bio-evolutionary uncertainty, as Deleuze might suggest. Kristeva writes that 'by way of abjection, primitive societies have marked out a precise area of their culture in order to separate it from the threatening world of animals or animalism'. Disgust, she says, is a 'safeguard' against the incursion of the animality, of chaos, and of irrationality; the ambivalence of the abject, 'the primers of my culture'.[62] If abject art has the capacity to wound, therefore, it also in doing so has the capacity to shore up the human against its own latent animality. Here, the connection implicit in Freud between the human-animal divide and the subject-object is made more explicit. The institution of the subject's sense of self is analogous (and in some societies, Kristeva suggests, homologous) to the establishment of the human as species being.

We may find this connection too in the work of Elizabeth Grosz who argues that the dividing line between human and animal — between a rational subject and the savage beast — is often extended metonymically to women, children, slaves, and other marginal identities, as 'the alignment of the most abjected others with animals is almost ubiquitous'.[63] The next chapter, on two of Thomas Hardy's 'Darwinian' fictions, will explore this notion further. For now, however, it might suffice to observe that Kristeva's work troubles Deleuze's desire to read in that of Zola a revelation of transformative potential, particularly in the context of his fiction's apparent concern with social-Darwinian conflict and biological determination. For that desire — to locate in our disgust at Zola's work a force that transcends both the animal and the human — may in fact conceal the very resistance to Darwin's Copernican revolution with which Freud was so concerned in his *Introductory Lectures*.

Conclusion

At the outset of the Conclusion to *On the Origin of Species*, Darwin defines the book's rhetorical structure and his elaboration of the theory of natural selection as 'one long argument' (*Origin*: 459). It is intended as a way of acknowledging the provisionality of his work, as well as the rhetorical structure of the text, which puts forward a multiplicity of facts and deductions in the service of a single idea, natural selection, answering possible weaknesses and disagreements as he does so. This phrase, however, and its attempt at rhetorical hedging, seems to register on another level the agonistic character of life the theory seemed to propose, as well as the manner in which such an explosively materialist and pugilistic account of human relations would generate further arguments in fields well outside the

62 Ibid., p. 2.
63 Grosz, *Becoming Undone*, p. 12.

range of evolutionary biology. The work of Émile Zola, despite the complexity of Darwin's position in the novelist's *œuvre*, seems ironically to encompass this coming together of the discursive and the biological in its own narrative arguments on biological determination and social antagonism. This is achieved most explicitly in *Germinal*, which stages an argument between seemingly opposed conceptions of political possibility in response to the social thesis of natural selection: undergirding the dialectic of Étienne's naïve progressivist Lamarckian natural selection and Souvarine's nihilistic embrace of a Darwinian death drive is a common acceptance of struggle as an ontological principle of social, political, and biological life. Yet, in doing so, this novel helps to chart not only the historical and intellectual trajectories of natural selection's social and political interpretations in the French nineteenth century, but signals Zola's awareness of these trajectories as well as of the conditions of reception that mediated his (and others') engagement in his time. This same auto-critical gesture is repeated in implicit ways in *L'Œuvre* and *Le Ventre de Paris*, both of which, like *Germinal*, use the figure of the failed revolutionary through which to outline a theory of the social Darwinian and hereditary conditions of life in Second Empire Paris, as well as fantasies (political, aesthetic) of transcending those conditions.

Yet, for all that Zola's work implicitly questions the discursive conditions of its own scientific objectivity and, by extension, a socially deterministic conception of natural selection, these novels nevertheless do seem to offer a pitiless image of life determined by social Darwinian competition and hereditary fate. Each failed revolutionary — Étienne, Florent, and Claude — in Zola's work seems to index a wish to transcend the social and biological struggle for existence represented by their historical milieu, as well as its impossibility, as political and aesthetic projects of socio-biological transformation are met with biological necessity, exile, and death. If the political failures of Florent and Étienne seem to imply that heredity and struggle always win out over human agency and political radicalism, the crushing of the aesthetic vision for a radically transformative realism, as represented by Claude (and his companion, Sandoz), seems more definitively to impugn Zola's own project of the experimental novel and its pretensions to social reform.

We may, however, interpret this auto-critical gesture in less conclusive terms. Zola's deflation of political and aesthetic possibility contra biological determination might instead be understood as a more contingent and specific critique of messianic revolutionary thinking which itself relies on a deterministic or teleological theory of change. For Deleuze, Zola's fiction does not promise a predetermined revolutionary trajectory, but rather confirms in its juxtaposition of narrative levels, and in its depiction of violence, the immanently revolutionary nature of biological life itself. Such an argument inverts and affirms the argument made by Souvarine in *Germinal* that a social and political ontology traversed by constant biological transformation would mitigate against historical certainty by suggesting that precisely this absence of certainty is revolutionary change's condition of possibility. Kristeva, while also affirming the potentially transformative character of violence in Zola's work, also suggests that by too-readily embracing its perverse allures we might consolidate rather than wound the very egoic defence upon which a conception of human exceptionality subsists.

If Zola's work fails to resolve this tension between competing conceptions of post-Darwinian humanity, that it stages such a critical and theoretical antimony in its formal rhetoric and thematic concerns is itself a notable conclusion. Contrary to much critical opinion, Zola's work addresses the problems of its own purported objectivity and the tensions that arise in an attempt to have literary art and politics engage with the biological destiny of humanity, and even seek to transform it. That these disputations only produce further aporia is testimony to these novels' ability to accommodate complexity, and to mobilise irresolution against the stifling epistemological certainty of evolutionary determinism, even as Zola's narratives seem to reconfirm the pessimistic nature of biological destiny.

CHAPTER 2

❖

Thomas Hardy: 'Relations between the Sexes'

> With respect to female birds feeling a preference for particular males, we must bear in mind that we can judge of choice being exerted, only by placing ourselves in imagination in the same position. If an inhabitant of another planet were to behold a number of young rustics at a fair, courting and quarrelling over a pretty girl, like birds at one of their places of assemblage, he would be able to infer that she had the power of choice only by observing the eagerness of the wooers to please her, and to display their finery. — CHARLES DARWIN (*Descent*: II, 122)
>
> Life being a physiological fact, its honest portrayal must be largely concerned with, for one thing, the relations of the sexes. — THOMAS HARDY[1]

As a matter of historical and literary fact, that Thomas Hardy's fiction and poetry was influenced by the scientific work of Charles Darwin appears to need little further confirmation. For not only was Hardy's literary 'vision of things' typical of 'a world view widely present in the late nineteenth century' in which both evolutionary and specifically Darwinian thinking featured prominently, there is also compelling evidence to suggest that Hardy's engagement with Darwin went beyond influence by cultural osmosis.[2] Florence Hardy's biography asserts that her husband was in fact 'one of the earliest acclaimers of Darwin', a claim that seems to be supported by the fact that Hardy was one of those in attendance at Darwin's funeral in Westminster Abbey in 1888.[3] He appears to confirm this idea himself in a letter to the American scholar Helen Garwood. Enumerating his non-literary influences, Hardy offers a constellation of canonical scientific and philosophical writers from the nineteenth century, citing Darwin's name first, among a list of other major scientific, political, and social theorists of his time: '[my] pages show harmony of view with Darwin, Huxley, Spencer, Comte, Hume, Mill, and others'.[4]

1 Thomas Hardy, 'Candour in English Fiction', in *Thomas Hardy's Public Voice*, ed. by Millgate, pp. 95–102 (p. 95).

2 J. Hillis Miller, *Thomas Hardy: Distance and Desire* (Belknap Press of Harvard University Press, 1970), pp. 16–17.

3 Florence Emily Hardy, *The Early Life of Thomas Hardy, 1840–1891* (Cambridge University Press, 2011), p. 198.

4 Originally cited in Helen Garwood, *Thomas Hardy: An Illustration of the Philosophy of Schopenhauer* (John C. Winston, 1911); Carl Jefferson Weber, *Hardy of Wessex: His Life and Literary Career* (Columbia University Press, 1940), p. 163.

In a more explicitly literary context, Hardy, in the prefatory 'Apology' for his *Late Lyrics*, responds to the alleged pessimism of his work and describes his poetry as a form of 'evolutionary meliorism'. This term might be indexed to a concern with gradual transformation as such, but it cannot be dissociated from the name of Darwin or from biological thought in the context of Hardy's stated influences or the wider intellectual culture of British culture at the turn of the century.[5]

Unlike the example of Zola, then, although Hardy's erudition is known to have been catholic and wide-ranging, the empirical claim of a genealogical and intellectual connection between Hardy and Darwin seems, comparatively, historically and discursively unproblematic.[6] If Zola's 'reading' of Darwin needs to be understood in the context of the latter's mediation and transformation through translation, political interpretation, and its passage into the realm of wider cultural and political assumptions, Hardy's encounter with Darwin seems culturally, linguistically, and historically immediate by contrast. Yet, there is more in common between Zola's and Hardy's respective receptions of Darwin than appears at first glance. For if Zola's engagement with Darwin's theory of natural selection is as an explicitly political and paradoxically Lamarckian concept of social evolution, Hardy too (as is indicated by his letter to Garwood) read Darwinian evolution as part of or in relation to evolutionary social theory. Hardy, we know, made notes on Francis Galton's foundational work of eugenic theory, *Inquiries into the Human Faculty and its Development* (1883) and attended a meeting of the Eugenics Education Society in 1910.[7] The notebooks too contain numerous references to the works of Herbert Spencer whose influence on Hardy, although 'not easy to particularize' because of the wide-ranging nature of its subject matter, includes passages from Darwin quoted by Spencer in his *Principles of Biology*, in which Spencer makes the case for the 'survival of the fittest', as well as the primacy of the individual.[8]

In that light, Hardy's 'evolutionary meliorism', if it is to be taken seriously as an indication of a Darwinian approach to literary creation, like Zola's quasi-social-Darwinian Naturalism, is an outlook that conjoins dourly materialist realism with social evolutionary optimism. In the *Late Lyrics* preface, Hardy refers to his poem 'In Tenebris II' as a way of characterising its double perspective: 'if way to the Better there be, it exacts a full look at the Worst'.[9] This observation could well apply to the way Zola's work, in the name of social melioration, seems to entail a confrontation with hereditary determinism, the struggle for existence, and their morbid consequences. The eugenic and social-Darwinian view of the nineteenth century also entailed a similar dynamic between knowledge and progress. For

5 Thomas Hardy, *Late Lyrics and Earlier* (Macmillan, 1922), p. xiii.

6 For a demonstration of Hardy's reading, see his *Literary Notebooks*, which include notes, quotations, and commentary on topics as diverse as Ancient Greek tragedy, Fourier's theory of the passions, the natural history of sea-life, and the concept of historical progress: *The Literary Notebooks of Thomas Hardy*, ed. by Lennart A. Björk, 2 vols (New York University Press, 1985).

7 Angelique Richardson, 'Hardy and Biology', in *Thomas Hardy: Texts and Contexts*, ed. by Phillip Mallet (Palgrave Macmillan, 2002), pp. 156–79 (p. 158).

8 Lennart A. Björk, 'Annotation 980', in *The Literary Notebooks of Thomas Hardy*, ed. by Björk, I, p. 341; Herbert Spencer, *The Principles of Biology* (William & Norgate, 1864), p. 445.

9 Hardy, *Late Lyrics and Earlier*, p. viii.

a eugenicist like Galton, understanding precisely the nature and distribution of hereditary weaknesses of 'human stock' was necessary to ensure that practices such as selective breeding could be employed to improve human society. Spencer's view is distinctive in its Lamarckian belief in social progress but this too, in insisting on the necessity of social and economic competition, combines a pessimistic realism with evolutionary hope. However questionable Galton's and Spencer's conceptions of 'Better', and however ideologically tendentious their interpretation of Darwin, each understands a biologically-grounded encounter with all that is considered 'the Worst' as a necessary precondition for optimism and a movement towards biological and therefore social amelioration.

This parallel in Hardy's reception of Darwin with Zola's creative social Darwinism complicates our understanding of the significance of Darwinism in Hardy's fiction. Known primarily for its focus on 'the Worst' — including tragic death, failure, suicide, infanticide, rape — and a wider sense of the arbitrary, deterministic nature of such events, the apparent pessimism of Hardy's novels is often credited to the stimulus of a foundationally Darwinian *Weltanschauung*. For if one early critic noted Hardy's preoccupation, 'like a naturalist with a bone', with apparently neutral (if not exclusively Darwinian) evolutionary thematic concerns, like deep time and genealogy, later critics specifically connected Hardy's pessimism with a reading of Darwin, which emphasised the loss of human agency, the contingency of human life, and determinism in his fiction.[10] Gillian Beer argues that 'plot in Hardy is almost always tragic or malign: it involves the overthrow of the individual either by the inevitability of death or by the machinations (or disregard) of "crass casualty", a reference to Hardy's poem 'Hap', which laments that human suffering comes not from a malign deity but from the indifference of a materialist universe and the passage of time.[11]

Other critics working on Hardy's Darwinism have rejected the notion that tragedy and pessimism represent the horizon of Hardy's evolutionary thought, as well as the obverse claim — partially suggested by Hardy himself — that his work was optimistically socially-Darwinian. George Levine elaborates the way Hardy's works register and instantiate a materialist form of 'enchantment', according to which Hardy's tragic materialism is supplemented by and in tension with a parallel concern with 'fullness, plenitude, or liveliness'; for Levine, the outline of this vision can be found in Darwin's 'feeling and valuing' in Darwin's thought as well as the importance it places on sexuality.[12] This 'romantic materialism', it has been argued, can be traced back to Darwin's reading of Romantics, like Milton.[13] Its most

10 Lionel Johnson, *The Art of Thomas Hardy* (Mathews & Lane, 1894), p. 65; Ross Shideler, *Questioning the Father: From Darwin to Zola, Ibsen, Strindberg, and Hardy* (Stanford University Press, 1999); Roger Ebbatson, *The Evolutionary Self: Hardy, Forster, Lawrence* (Rowman & Littlefield, 1982); David Lodge, '"The Woodlanders": A Darwinian Pastoral Elegy', in *Working with Structuralism: Essays and Reviews on Nineteenth- and Twentieth-Century Literature* (Routledge, 1981), pp. 79–94.

11 Beer, *Darwin's Plots*, p. 223.

12 George Levine, *Darwin Loves You: Natural Selection and the Re-Enchantment of the World* (Princeton University Press, 2008), p. 169.

13 Beer, *Darwin's Plots*, p. 37.

famous example is the image of the 'entangled bank' with which Darwin concludes *On the Origin of Species*:

> It is interesting to contemplate an entangled bank, clothed with many plants of many kinds, with birds singing on the bushes, with various insects flitting about, and with worms crawling through the damp earth, and to reflect that these elaborately constructed forms, so different from each other, and dependent on each other in so complex a manner, have all been produced by laws acting around us. [...]
>
> There is grandeur in this view of life, with its several powers, having been originally breathed by the Creator into a few forms or into one; and that, whilst this planet has gone cycling on according to the fixed law of gravity, from so simple a beginning endless forms most beautiful and most wonderful have been, and are being, evolved. (*Origin*: 490)

Entanglement has become an increasingly important signifier in recent attempts to theorise human/non-human relations. For Jane Bennett, an encounter with a 'vital materiality [...] draws human attention sideways, away from an ontologically ranked Great Chain of Being and toward a greater appreciation of the complex entanglements of humans and nonhumans'.[14] Drawing on the work of Gilles Deleuze and Félix Guattari, Bennett asks us to 'picture an ontological field without any unequivocal demarcations between human, animal, vegetable, or mineral', postulating a 'monist' or 'flat' ontology populated by a multiplicity in which humans, as Darwin suggests, are no longer at the centre.[15] Other conceptions of entanglement, part of a wider theoretical 'return to materiality', draw on fields as diverse as quantum physics, trans-species companionship, and literary Romanticism to theorise, indirectly, the problem of relationality that Darwin raises in *On the Origin of Species*.[16] Darwin's own entangled bank seems to offer a radical ontological picture and postulates forms of complex relationality. Combining the sublime and the pastoral, the global and the local, it articulates a field of organic being in which the subjective, affective, and creative seem to exist in dynamic relationship with cyclical repetition and determinate laws of nature. Both widening and narrowing the social-Darwinian aperture, it focuses on the complex and subterranean minutiae of non-human life and its wider imbrication in a network of infinite relationality.

This interpretation of Darwin's scientific romanticism might, in turn, help us articulate a reading of Hardy's Darwinian influence beyond a binary of irrevocable pessimism or Whiggish optimism, to focus instead on his work's representation of different forms of entangled relations. This chapter will examine not only the way in which Hardy's novels figure human relations with the natural world (and thereby the Darwinian view of nature, with which Hardy was clearly concerned), but is also concerned with other forms of entanglement, which are often thought

14 Jane Bennett, *Vibrant Matter: A Political Ecology of Things* (Duke University Press, 2010), p. 112.

15 Ibid., p. 117.

16 Karen Barad, *Meeting the Universe Halfway: Quantum Physics and the Entanglement of Matter and Meaning* (Duke University Press, 2007); Timothy Morton, *Hyperobjects: Philosophy and Ecology after the End of the World* (University of Minnesota Press, 2013); Donna J. Haraway, *The Companion Species Manifesto: Dogs, People and Significant Otherness by Haraway* (University of Chicago Press, 2003).

to be specifically human forms of mutual relation: love and the erotic. This dual focus articulates a tension between the apparent pessimism of his fiction's vision of humanity's relation with its evolutionary predicament, and the optimism of his work in relation to the creative and subversive possibilities of erotic relation. Of course, such a tension is not readily apparent in other critical studies, since love in Hardy is not commonly associated with optimism. As Jane Thomas has argued, drawing on a Lacanian conception of desire and the impossibility of its satisfaction, love relations are intrinsic to the tragic vision of Hardy's novels:

> Hardy's lovers, male and female, become fascinated by the beloved other, driven by the prospect of pleasurable possession to assuage the yearning at the heart of their being. Their desire is intensified in exact proportion to the unavailability of the beloved object, whilst possession (a rare phenomenon in Hardy) results not in fulfilment but, at best, indifference and, at worst, bitter disappointment and the troubling resurgence of desire for someone or something else.[17]

Here, however, I read the relationship between love and tragedy in Hardy's work as reflecting the tension in Darwin's work, in which the characterisation of sexual relationships and desire seems to contradict that of natural selection. The conception of a specifically evolutionary sexual relationship in Darwin first appears in *On the Origin of Species*, which proposes that the expression of sexual difference in animals sexual depends 'not on a struggle for existence' but on a comparatively 'less rigorous' competition to attract potential mates and for those mates, in turn, to be attracted by the most suitable suitors (*Origin*: 88). If in Zola, critical discourse, contemplation, or even art offers an interval in its own biologically deterministic narrative logic, Darwin's reference to sexual competition opens up a comparable space within the matrix of natural selection. Certain features of life — namely the characteristic features of sexual dimorphism — seem to evade the rigours of the struggle for life and depend instead on a comparatively benign form of sexual agonism. If 'sexual relation' in the Lacanian conception of the love plot always results in desire's disappointment, we might nevertheless point out that bathos is often preferable to extinction.[18]

This suggestive proposal in Darwin's *On the Origin of Species* anticipates of course his other major theory of evolutionary change, sexual selection, which is introduced twelve years later in *The Descent of Man*. In the former, Darwin postulates that individuals compete to survive and therefore to propagate the species, with evolutionary selection therefore favouring individuals who develop and reproduce morphological traits or behavioural strategies that aid them in the 'struggle for life'. By contrast, sexual selection relates to the dynamics of reproduction, understood as distinct from the pressures of survival, and involves competition between individuals (often, in Darwin's work, the male) to attract mating partners, who also exercise some choice in their selection of mates. This process of selection favours

17 Jane Thomas, *Thomas Hardy and Desire: Conceptions of the Self* (Palgrave Macmillan, 2013), p. 72.

18 In *Seminar XVII* Lacan famously states that sexuality is defined, in fact, by non-relation, claiming that 'What the master's discourse uncovers is that there is no sexual relation': Jacques Lacan, *The Seminar of Jacques Lacan, Book XVII: The Other Side of Psychoanalysis*, trans. by Russell Grigg (Norton, 2007).

seemingly unnecessary features of certain species that are represented by 'courage and pugnacity — their ornaments of many kinds — their organs for producing vocal or instrumental music — and their glands for emitting odours; most of these latter structures serving only to allure or excite' (*Descent*: I, 257–58). Darwin, while discussing sexual selection by birds (as quoted in this chapter's epigraph), explicitly acknowledges that we find analogous behaviours in human relations. Human rituals, such as the 'courting and quarrelling over a pretty girl', he writes, not only allow us to see animal life as engaging in the same kinds of superfluous behaviour, but emphasise the complex relations of desire between mating couples. If the male is ordinarily supposed to be active while the female is passive, the 'power of choice' in fact lies with the female of the species insofar as she observes and selects her mate based on his 'eagerness' and 'finery' (*Descent*: II, 122). Unlike natural selection, therefore, sexual competition is neither fatal nor wholly mechanistic but depends on subjective sensual experience, observation, and a dialectic of desire. Such desire produces excessive bodily morphologies and ritual behaviours such as dancing or song whose apparent final cause is the reproduction of the species but whose proximate cause is paradoxically self-sufficient by serving only to incite (sometimes homosexual) pleasure and desire.[19] If in natural selection Darwin can locate the biological expression of Malthusian necessity, it is in the dynamics of sexual competition and attraction that he speculatively locates the origin of aesthetic and erotic phenomena that exceed the instrumental logic of their origins: music, self-decoration, ornamentality, performance, all of which function as ways of instantiating and mediating sexual desire and its relation to the other.

How might this framing of Darwinian evolution, which privileges the later theorisation of sexuality in *The Descent of Man* over the dynamics of natural selection, reorient our understanding of Hardy's writing and its response to Darwin in particular? How might it alter our understanding of the way in which Hardy's work is thought to relate to the question of the post-Darwinian human? If Bert Bender's work on sexual selection in American fiction at the turn of the nineteenth century suggests that its integration into narratives of courtship produced an increasingly dark and violent conception of sexuality, Travis Landry's study of its influence on Spanish fiction reframes Darwinian sexual selection in terms of female choice and indeterminate forms of change.[20] For J. Hillis Miller, in his study of desire in Hardy's novels, the recuperative potential in these narratives can be understood in relation to Arthur Schopenhauer's conception of aesthetic liberation (*Befreiung*), which lies in his characters' refusal of the meaningless, violent determinism in both the 'yielding to the magical power of music and falling in love'.[21] If for Darwin, love and music, the sexual and the aesthetic, are linked, for Hillis Miller they might form the radical or reparative core of Hardy's engagement

19 Donovan O. Schaefer, 'Darwin's Orchids: Evolution, Natural Law, and the Diversity of Desire', *GLQ: A Journal of Lesbian and Gay Studies*, 27.4 (2021), pp. 525–50, doi:10.1215/10642684-9316824.

20 Bert Bender, *The Descent of Love: Darwin and the Theory of Sexual Selection in American Fiction, 1871–1926* (University of Pennsylvania Press, 1996); Travis Landry, *Subversive Seduction: Darwin, Sexual Selection, and the Spanish Novel* (University of Washington Press, 2013), pp. 47–69.

21 Miller, *Thomas Hardy*, p. 23.

with evolution. Darwinian natural selection is associated in Zola's work with brutality, violence, and the impossibility of social amelioration. But sexual selection can be seen as a source, for Hardy, through which to develop a dramatisation of human relations which refuses pessimistic determinism while seeming to remain Darwinian in character (or altering what we take to mean by the term *Darwinian*).

Angelique Richardson has established how the advent of Darwinian sexual selection meant that, by 1872, as Hardy's career as a novelist began to take shape, 'love and human relations were [...] at the forefront of evolutionary science' and 'the love-plot [...] was now the focus of evolutionary narrative'.[22] Hardy's essayistic writing, too, invites us to see his evolutionary narratives as concerned with 'life [as] a physiological fact' as a result of which 'its honest portrayal must be largely concerned with, for one thing, the relations between the sexes'.[23] The representation of these relations, Hardy suggests elsewhere, had up to that point in history been limited by the necessity to conform to 'social forms and ordinances'; now, 'the 'position of men and women in nature [...] may be taken up and treated frankly'.[24] If Zola's realism is concerned with extremities of representation that reflect the violence and determinism of natural (and social) selection, for Hardy, 'truth' and frankness lie in an 'honest' and therefore scientific representation of sexuality, reflecting the influence of Darwin's later work, *The Descent of Man*, and its theory of sexual selection. Hardy's understanding of his own narrative apparatus, so frequently concerned with dramatising love plots, courtship, and their failure, promises to offer a distinctive articulation of post-Darwinian human relations which refuses the social-Darwinian combination of pessimism and deterministic renunciation in the service of a heedless and violent social optimism.

While this understanding of Hardy's reception of Darwin might countervail the most pessimistic conceptions of Hardy's engagement with Darwin, it does not easily distance his fiction from biologically essentialist or determinist conceptions of human life, sex/gender, and sexual desire. The pioneering naturalist, Antoinette Brown Blackwell, noted as early as 1875 in *The Sexes throughout Nature* that Darwin's narrative of sexual selection seems to reify what she called even then the 'time-honoured' concept of the male as normal and powerful in relation to the aberrant and passive nature of the female.[25] This proto-feminist view anticipates later philosophical work that challenges the essential connection between biological sex and gender. For Simone de Beauvoir, by 1949, 'one is not born but becomes a woman' and feminist thought should 'repudiate any frame of reference that presupposes the existence of a natural hierarchy of values — for example, that of an evolutionary hierarchy', including whether 'the female body is more infantile than the male, if it is closer to or further from that of the higher primates'.[26] If Hardy's work reflects a 'new, biologized, romance that was coming to birth' in

22 Richardson, 'Hardy and Biology', p. 157.

23 Hardy, 'Candour in English Fiction', p. 95.

24 Raymond Blathwayt, 'A Chat with the Author of Tess', *Black and White*, 27 August 1892, p. 240.

25 Antoinette Brown Blackwell, *The Sexes throughout Nature* (Hyperion Press, 1976), p. 122.

26 Simone de Beauvoir, *The Second Sex*, ed. by Sheila Malovany-Chevallier, trans. by Constance Borde (Vintage Books, 2011), p. 131.

response to sexual selection, it also seems to reflect the patriarchal biologism that these thinkers critique, by drawing on many conventional signifiers of male sexual prowess and feminine beauty and implicitly locating them in biological conceptions of sexuality.[27] His early work, which forms part of the analysis of this chapter, *A Pair of Blue Eyes*, abounds in cliches of masculinity such as strong facial hair and displays of social dominance, as well as feminine, childishness, blushing, and a love of 'adornment in women' which function as integral parts of Darwin's conception of sexual competition and courtship.[28] Therefore, if a focus on sexuality in Hardy offers an alternative to reading his work in relation to Darwinian pessimism and natural selection, it still does not offer us a royal road to Darwinian *Befreiung*. Rather, such a potentiality should be tempered by an understanding of the imbrication of Darwinism in a historical project of biological sexism.

If in Zola, humanity becomes biologically abject as a form of life that is mechanistically determined by social and biological conditions, to what extent, if at all, does Hardy's engagement with Darwin and sexual selection offer a different view of human possibility and creativity? In Zola's fiction, a qualified form of human agency can be found in forms of critical reflection, art, or the revolutionary act (if not its success), even as humanity seems locked within the iron cage of its own egoic self-conception (if not an absolute form of biological essentialism). What form of human possibility can be found in Hardy's representation of the dynamics of Darwinian sexual relations, which in its focus on attraction and desire offers a different perspective on human relations and agency — only to restrict it to the phallic activity of the sexually predatory male?

This chapter addresses these questions by reading two of Hardy's novels, the early work *A Pair of Blue Eyes* (1872), which forms a part of Angelique Richardson's important analysis of the Hardy-Darwin connection, and his later, better-known novel, *The Return of the Native* (1878). It begins by reading in these two novels an affirmation of the possibility (or, in the terms of Hardy's vision of 'the Worst', the ethical necessity) of reading these novels' narratives as an affirmation of a cruel and deterministic post-Darwinian humanity. In the sections that follow, the chapter goes on to investigate further whether and to what extent Hardy's engagement with Darwin in these works — in this case, through the lens of the theory of sexual selection, as opposed to natural selection — furnishes us with a vision of human agency counter to the pessimism with which Hardy's Darwinism is associated. To do so, it explores Hardy's representation of games, folk ritual, and dance, examining the way these formalised modes of relation codify forms of sexual interest and competition and, in doing so, stage a biologically grounded conception of human life and sexuality that both reinforces and challenges Victorian conceptions of gender performance and sexuality. The final sections of this chapter return to more

27 Angelique Richardson, '"How I Mismated Myself for Love of You!": The Biologization of Romance in Hardy's "A Group of Noble Dames"', *Thomas Hardy Journal*, 14.2 (1998), pp. 59–76 (p. 76).

28 Angelique Richardson, '"Some Science Underlies All Art": The Dramatization of Sexual Selection and Racial Biology in Thomas Hardy's *A Pair of Blue Eyes* and *The Well-Beloved*', *Journal of Victorian Culture*, 3.2 (1998), pp. 302–38, doi:10.1080/13555509809505967.

recent theorisation of sexual selection, particularly in the work of Elizabeth Grosz, who elaborates conception of evolution as an engine of radical and transformative difference. Grosz sees in Darwinian sexual selection a potentially subversive conception of sexuality, grounded in biological materiality, and argues for a renewed understanding of the radically excessive and subversively creative nature of biological sexuation. In this light, re-reading the way folk ritual and dance are figured in Hardy's works, I explore how Grosz's conception of sexual difference and sexual selection allows us to see Hardy's Darwin — and therefore Darwin's work as such — as containing, perhaps despite itself, the kernel of an anti-determinist conception of *Befreiung*.

The Failure of Evolutionary Meliorism

On the face of things, *A Pair of Blue Eyes* and *The Return of the Native* elaborate tragically deterministic narratives that mirror the causal logic of biological necessity. The earlier novel, offers the story of Elfride Swancourt, the semi-aristocratic daughter of a rector in a country parish in fictional Lower Wessex, and of her courtship by a series of admirers: an apprentice architect named Stephen Smith; Smith's mentor, the urbane amateur geologist and writer, Henry Knight; and a local, widowed aristocratic peer of the realm, Lord Luxellian. The plot hinges on Elfride's father's rejection of Smith's advances, upon discovering Smith's inferior social status, and on Henry Knight's later rejection of Elfride, upon discovering the existence of her previous courtship, though he remains ignorant of the suitor's identity. Finally, after a period of illness precipitated by the social shame of her two failed relationships, Elfride marries Luxellian out of social and economic convenience, becoming a surrogate mother to Luxellian's children. She dies tragically, in Hardy's words, 'with a miscarriage' soon after (*Blue Eyes*: 353).[29] The novel's concluding scene is suggestively similar to that of Zola's *L'Œuvre*. Elfride's dead body is transported by train back to Wessex as, on the very same train, Smith and Knight quarrel over their competing claims to Elfride's hand, until they jointly discover the news of Elfride's demise.

The Darwinian subtext of the plot may not be immediately apparent, but the homology with Zola's symbol of the locomotive is an instructive place to begin. Just as the onrushing train and its conjunction with Claude's funeral rites in *L'Œuvre* signals the inexorable (evolutionary) movement of heredity and a propulsion towards death, here the image of Elfride's coffin being carried to her resting-place in Wessex may symbolise the fixity of biological fate. The pathos of this conclusion lies in the sense that for all their collective attempts to control their respective romantic lives, which forms the narrative of the novel's 'love-plot', a deeper, fatal logic over which neither Elfride nor her suitors have any control operates beneath

29 In the original, serialised version of the novel, Hardy offered no explanation for her death; the almost equally ambivalent explanation of 'miscarriage' was a later addition to the Wessex Edition of the text which serves as the basis for the standard edition of this novel (*Blue Eyes*: 374, n. 353). For a review of the editorial evolution of this novel see Alan Manford's notes on the text (*Blue Eyes*: xxxix–xliv).

their intentional actions. This conjunction of plot and its contingencies (failed courtships, social missteps, accidental events, and so on) and narrative movement towards death as represented by the locomotive, as in Zola's fiction, stages the Deleuzean two-tiered narrative of heredity. Nature in Hardy is 'watchful', it 'picks out, and favours each successful competitor, but rejects failures'.[30] Elfride fails not only because she is unable — in the eyes of her father and her second lover, avatars of a wider cultural set of patriarchal expectations — to negotiate rigid gendered and societal and sexual mores by becoming romantically entangled with notionally unsuitable mates. The implication of her premature death in childbirth marks her out and her miscarried child, like Claude and his son in *L'Œuvre*, as a member of the species of the biologically weak.

Here in Wessex, as in Zola's Paris, the social and the biological are intertwined: social marginalisation, poverty, transgressive sexuality, and political or aesthetic radicalism are overlaid with biological fate, one confirming the other. In *A Pair of Blue Eyes*, this convergence of the social and the biological can be read in the ambiguous term, *miscarriage*, used to describe the circumstances of Elfride's death. The term's polysemy, of course, raises the spectre of infant mortality, which we find in Hardy's *Jude the Obscure* (1895). However, the putative presence or absence of a child is merely implicit: the word is used in the text only to refer to Elfride's death, whose pregnancy we must infer. The term can be both associated with Elfride's death, suggesting not only a biological flaw or weakness (a 'miscarriage'), as well as a deviation from a social track or telos (as 'mis-carriage') that would otherwise have guaranteed her health. That her death might represent a failure of natural justice is also suggested here, but such subtextual ethical concerns seems to pale in the novel's sensationalist ending.

Hardy's later novel *The Return of the Native* mirrors *A Pair of Blue Eyes*, offering a love-plot that involves a set of male competitors all focused on a central female figure, Eustacia Vye, whose reward for being the object of desire, like Elfride, is death. Published serially in the *Belgravia* magazine in 1878, six years after the publication of *The Descent of Man*, the novel opens with a panoramic view of the 'the vast tract of unenclosed wild known as Egdon Heath', in Hardy's fictional Wessex, an ancient, mysterious, and seemingly timeless country where 'civilisation was its enemy'. This landscape — 'a heathy, furzy, briary wilderness' — is the landscape against which we are introduced to the novel's central female figure, Eustacia Vye, 'the raw material of a divinity', a dark-haired woman whose passions made her the model of a deity but 'not quite a model woman'. Her narrative in the novel begins at the conclusion of a brief affair with a local man named Damon Wildeve and as Eustacia finds herself infatuated with the novel's titular native, Clym Yeobright, a successful businessman returned home to Egdon from Paris with the ambition to establish a school in which the inhabitants of the heath could be educated (*Return*: 47). After marrying, the two become disillusioned; Eustacia's love for Clym and her ambitions to leave Egdon for Paris are thwarted, first, by Clym's educationalist ambitions and desire to remain in Wessex and, second, by a sudden deterioration

30 Mark Asquith, *Thomas Hardy, Metaphysics and Music* (Springer, 2005), p. 35.

in his eyesight and his ensuing social isolation. As a result, Eustacia's interest in Wildeve is reignited but, as Wildeve waits to liaise with Eustacia with the intention of eloping to Paris, Eustacia, terrified of the prospect of her plan becoming exposed, throws herself, or falls, into a nearby weir and drowns.[31]

At the level of narrative, this novel mirrors the ambivalence of the earlier novel by dramatising the central dynamic of sexual selection, in its presentation of a heteronormatively orthodox love-plot (two men compete over a single female object of desire), but through this offering little in the way of agential power for any of its participants. As in *A Pair of Blue Eyes*, the tragic death of the novel's female protagonist dramatises the extent to which fate in the narrative functions according to a remorseless logic of determination that exceeds human agency. And indeed, if sexual selection affords the female sex the 'power of choice', choice itself is represented as malign. In Eustacia's death we can again see the conjunction of fate and contingency. Whether read as suicide or accident, Eustacia's end appears narratively contingent, not predetermined: we are led to believe that the fatal decision to cross the heath at night and elope with Wildeve may have been avoided had a letter, written by Clym in an attempt at reconciliation, been delivered moments earlier. Yet, as Clym pursues Eustacia into the darkness, what he finds is a force that instantiates a remorseless, anti-civilisational inevitability: he sees an opaque 'vortex' of water in which 'a dark body was slowly borne by one of the backward currents' (*Return*: 378). If in Zola's works, humanity's relation to its own animality is figured in terms of social antagonisms — violence, starvation, insurrection — here the metaphor of 'submersion' defines the human's relation to nature, symbolised by the heath and its resistance to civilisation (as indexed by Paris). Each character is swallowed up by a 'hole' in the heath; Wildeve's body is recovered but all that can be found of Eustacia is her clothing. Like Elfride, Eustacia is victimised by her desire and that of her male competitors, Wildeve and Clym. The tragic logic that characterises the events may not be explicitly Darwinian, but is consistent with the wider diagnosis of Hardy's tragic vision which, like the plot of natural selection, is characterised by its indifference towards the fate of humanity. This view is echoed in critical studies of Hardy's plots of sexual courtship, which it has been argued are 'permeated by a nearly obsessive preoccupation with the effects of female choice' and its malign consequences. The 'sexual contract' found in sexual selection, Hardy's work implies, 'is used to grant women power with the one hand while taking it away with the other'.[32] The individual is overcome, as Miller puts it, by the 'measureless hollow' of unconcerned nature — one without creator or plan, in which the only certainty is death, particularly women who desire. In this novel,

31 The ambiguity surrounding Eustacia's death and the divergent readings it has prompted are discussed in Frank R. Giordano, 'Eustacia Vye's Suicide', *Texas Studies in Literature and Language*, 22.4 (1980), pp. 504–21. Regardless of *how* Eustacia dies, the sense of the inevitability of her death and her becoming 'swallowed up' remains consonant with the notion of Darwinian and materialist tragic fatalism.

32 Ruth Bernard Yeazell, 'Nature's Courtship Plot in Darwin and Ellis', *Yale Journal of Criticism*, 2.2 (1989), pp. 33–53; Nancy Armstrong, *Desire and Domestic Fiction: A Political History of the Novel* (Oxford University Press, 1987), p. 36.

as in *A Pair of Blue Eyes*, the plot of sexual selection is combined with the survivalist logic of natural selection, resulting in a particularly gendered form of tragedy.

At the level of plot, then, both these works seem to reproduce a fatalistic conception of human life according to which natural force has primacy over human agency, particularly that of women. Where Deleuze's reading of endings in Zola reveals how they are pregnant with renewal, the conclusions of these two novels by Hardy seem to offer little hope of either liberation or consolation for either sex. If, as Beer has argued, Hardy's focus on the death of an individual across a single generational span reflects a humanism opposed to the infinite space of evolutionary narrative, this also represents its limits, by seeming to ignore transgenerational possibilities of germinal transformation.[33] As D. H. Lawrence points out, Hardy's endings are doubly tragic because it is precisely those characters who embody or seek to attain a kind of autonomous individuality — contrary to the determinist account of Darwinian law — who are punished most severely. Recalling the dialectic of Deleuze's schema of life drive and death drive in Zola, Lawrence identifies and laments a narrow Darwinian logic of 'self-preservation', a conservative logic that reproduces itself in social contexts.[34] Hardy's 'aristocratic' characters, Lawrence suggests, such as Eustacia, Wildeve, and also Elfride, like Claude and Florent, are consigned to their tragic fates through their desire to repudiate the pre-determined social or biological laws of a 'struggle for existence', seeking instead to attain a 'fullness of being' outside the utility of mere survival.[35] As such, Hardy's Darwinian vision is analogous to Zola's insofar as evolution acts in a socially repressive manner, destroying those who seek to 'refuse' evolutionary destiny, those who struggle for something other than survival.

Escaping the Darwinian Abyss: Vision and Desire

The character of Hardy's pessimism — which I am here reading in relation to a wider Darwinian intellectual and cultural imaginary — is described by Miller in his study of Hardy's fiction. On the apparent failure of any kind of 'evolutionary meliorism', Miller writes that in Hardy's fiction:

> Man must endure things as they are. This endurance is made more painful by knowledge that if the Immanent Will does not come to consciousness the best man can hope for is that he will be lucky enough to 'darkle to extinction swift and sure'. The development of man is a mistake on the part of the vital energy of the earth. Man is not more fit for survival than the dinosaur or the sabre-toothed tiger.[36]

Unlike in Zola's work where a sense of conscious intelligence — in the form of critical reflection, political liberation, or art — offers the possibility (if not the

33 Beer, *Darwin's Plots*, p. 223.

34 D. H. Lawrence, 'Study of Thomas Hardy', in *Study of Thomas Hardy and Other Essays*, ed. by Bruce Steele (Cambridge University Press, 1985), pp. 3–132 (pp. 20–22).

35 Ibid., pp. 48–50.

36 Miller, *Thomas Hardy*, p. 16.

realisation) of liberation from 'things as they are', in Hardy's two novels it appears that a consciousness of an apparent biological imprisonment seems absent, taken up instead by the apparent narcissism of love and sexual desire. Zola's protagonists' lives (and deaths) are dedicated to these characters' attempts, both consciously and unconsciously, to liberate themselves and others from a stultifying or violent social and biological fate. Hardy's female protagonists, by contrast, if they can be understood to possess the Darwinian sexual 'power of choice', seem neither to possess an awareness of their own fate, nor any profound desire — beyond a general kind of Lawrencian counter-cultural individuality — to escape it. Eustacia's desire to escape from her home in Egdon Heath to Paris, the association with which is partly what attracts her to Clym, may stem from an unconscious desire to escape from nature to civilisation. However, in Eustacia's words, it is in the service of a desire for Parisian 'beauty' and 'charms', what Clym dismissively refers to as a 'longing for town bustle' (217). Elfride's power of choice, too, can be characterised as the capricious desire of a naïve young woman, whose willingness to be courted by (and ultimately reject) a man of such low social standing as Smith has ultimately tragic consequences. Whether this problematically gendered idea of these characters' desire to resist biological fate is warranted is a question we shall return to later in this chapter.

For now, however, another way of locating human agency in Hardy's Darwinian universe might be found in his novels' staging of vision.[37] Darwin's empiricism, argues Levine, is paradoxical because the observational position of a sovereign and rational subject, on which Darwin's work is based, is eventually problematised by the Darwinian conception of human-animal entanglement. This is another consequence of Darwin's Copernican revolution: the human, decentred and dethroned, can no longer claim a position of objective distance from nature. In *The Return of the Native* the problem of seeing is staged throughout the novel by its setting of the heath, which appears to us amid different kinds of darkness and visual obscurity. Characters in this work are seen dimly, or in silhouette, and the heath itself 'exhales' darkness. The act of lighting a torch — the signal Eustacia makes to her lover, Wildeve — is described as an act of 'Promethean rebelliousness' against 'foul times, cold darkness, misery and death': light as a movement against pessimism (58–59). If the cause of Eustacia's death is obscure, it is in part a function of the darkness in which it is shrouded; Clym, whose own blindness at this point has now healed, strains in the night to see into the 'hole' of the pool in which Eustacia eventually drowns and disappears. Darkness mitigates against a transparent kind of vision that can in turn be read as a reflection of Hardy's post-Darwinian pessimism and its displacement of transcendent observation.

However, darkness can also be understood to function symbolically in the novel as providing a form concealment through which a non-transcendent subject can find a form of agency. This is staged by Eustacia's position as she observes the object of her sexual desire, Clym, without his knowledge. Eustacia infiltrates a band of mummers which performs in Clym's home, obscuring both her identity and her

37 George Levine, 'The Perils of Observation', in *Darwin and the Novelists*, pp. 210–38.

gender behind her costume as, in turn, Clym observes her performance, unaware of her identity (or, as yet, her desire). Once the performance concludes, Eustacia finds another concealed position from which to observe her subject. In both cases, her observational removal is threatened by the possibility of recognition that is as alluring as it is disastrous. 'At moments during this performance,' the narrator observes, 'Eustacia was half in doubt about the security of her position; yet it had a fearful joy' (*Return*: 176). On the one hand, this position underlines the contingent nature of observational removal, which is here achieved not through a transcendence but concealment. On the other hand, this fragile position is afforded an additional affective charge in which the fear of being exposed is part of its perverse pleasure. For Eustacia, its function is not only to take up a position of a hidden observer, but also to be watched herself in the process, by the object of surveillance. This scene constructs a relationship of mutual engagement and desire, enabled and mediated by creative forms of self-concealment and misrecognition. Darkness may obscure the observation of a transcendent observer (or narrator), but it provides the conditions for different forms of relation for the characters in the novel, for whom intimate proximity is desired.

This construction of the hidden-in-plain-sight subject being observed unwittingly by her object problematises the structural conditions of a notionally realist objectivity and its vision of materialist life. Eustacia's position is not the view from nowhere of the rational empiricist observer, but one that enacts instead a contingent form of looking driven by a desire for and involvement with its object. Darwin himself revelled in being among nature rather than above it and casting off the fantasmic veil of transcendence required of empiricist surveillance. He describes in a letter, written soon before the publication of *On the Origin of Species*, how, after falling asleep in the grounds of the hydrotherapeutic clinic in Moor Park, he awakens surrounded by a 'chorus of birds singing [...] & squirrels running up trees & some Woodpeckers laughing'. Here, the desire for scientific mastery through empirical distance remains dormant, the constructed interval between observer and his subject is closed, and Darwin luxuriates in the freedom this grants him: 'I did not care one penny how any of the beasts or birds had been formed'.[38] Moreover, in the suspension of a constructed scientific objectivity, Darwin's interest in struggle as the causal vanishing point for the natural world shifts towards an appreciation of its aesthetic and affective character.

Perhaps, then, this movement between a desire for objective distance and a desire for intimate proximity is a possible source of an 'evolutionary meliorism': offering a form of active human agency without insisting on human transcendence, while also reframing our view from brutal antagonism towards aesthetic and affective states and relations. By claiming a standpoint, even through self-concealment, with which to view the objects of our scientific interest, while accepting its contingency, we may at once rediscover our position within the natural world (or our own animality) and begin to see something other than a relentless struggle to survive.

38 Charles Darwin, letter to Emma Darwin, 28 April 1858, in Darwin Correspondence Project <http://www.darwinproject.ac.uk/entry-2261> [accessed 3 November 2014].

To the extent that a pre-Darwinian conception of the human subject and species is premised on a position of disengaged objectivity in relation to the animal and nature, the challenge of the post-Darwinian human might be to find a position that remains grounded in animality, without becoming swallowed up.

The fortunes of Eustacia's object of surveillance and eventual husband, Clym Yeobright, allow us to explore this possibility further. In contrast to the gloomy barbarism of Egdon Heath, Clym sees himself as a rationalist and figure of enlightenment — the 'Rousseau of Egdon' — who 'had a conviction that the want of most men was knowledge of a sort that brings wisdom rather than affluence' (*Return*: 243). This conviction was nourished by Clym's knowledge of 'the central [...] thinkers of his date [which] he may have owed to his studious life in Paris, where he had become acquainted with ethical systems popular at the time' (*Return*: 204). Hardy is referring here, it seems likely, to Auguste Comte's theory of social progress, which states that humanity moves through successive periods of development leading ultimately to the 'positive stage' — a telos of total knowledge, the point at which 'the most exact and the most complete possible knowledge of the laws of nature' is attained.[39] The biological sciences are central for Comte: their extension into sociology is thought to provide the conditions for a rigorous science of humanity. Thus Clym's desire to found a school to educate the inhabitants of Egdon is part of a larger biologically informed, epistemological endeavour, whose origins are located in Paris, which has the completion of human knowledge at its core. In this respect, like Florent in *Le Ventre de Paris*, Clym possesses an ambitious if naïve conception of social liberation, which, like that of Sandoz in *L'Œuvre*, is premised on a project of positivistic knowledge. If in Zola's works, nothing less than revolution will do, for Clym a more reformist objective, to 'raise the class' of his fellow Egdonites, is the goal.

As part of his vision to 'raise the class' of his fellow Egdonites, Clym dedicates himself to more intense study and to 'read[ing] far into the small hours during many nights' (*Return*: 203; 270). However, the intensification of his effort is rewarded not with enlightenment (either for him or his notional pupils) but with darkness: he wakes one morning to find himself partially blinded and looking at the world as if through 'smoked glass': '[a]t every new attempt to look about him the same morbid sensibility to light was manifested, and excoriating tears ran down his cheeks' (270). Whatever the aetiology of this mysterious symptom, a symbolically freighted lack of sight functions for Clym, as it does in Darwin's letter, as a release from the painful isolation of scientific detachment. Indeed, paradoxically, it affords Clym the chance to move into closer proximity to the natural world, despite his lack of vision. In lieu of being able to study, linked metonymically to Parisian civilisation and positivist epistemology, Clym takes up furze-cutting, the local practice of harvesting for fuel the gorse bushes which cover the heath, and in doing so seems to melt into the

39 In 'Appendix C' of the Broadview Edition of *The Return of the Native*, Simon Avery writes that Hardy was 'a keen reader of [Comte's] work' (*Return*: 429), an argument borne out by Hardy's *Literary Notebooks* which contain numerous references to the French Positivist theorist; Auguste Comte, *Early Political Writings*, ed. by H. S. Jones (Cambridge University Press, 1998), p. 154.

natural territory in which he moves. Released from the agony of seeing (or reading), Clym becomes 'a brown spot in the midst of an expanse of olive-green gorse, and nothing more', taking pleasure in the forced limitation of his own predicament. He becomes the companion to 'creeped and winged things'; bees, grasshoppers, and 'huge flies' encircle him 'without knowing he was a man'; snakes and rabbits run about his feet, and '[n]one of them feared him' (273). In relinquishing the fantasy of scientific mastery, as embodied in his desire for enlightenment, Clym instead gains a different, more intimate kind of knowledge in a seeming relation of mutuality with the non-human environment. Like Zola's pathetic revolutionaries, the failure of Clym's social and scientific project of enlightenment seems to offer a different relation to the natural world — or in this case, the apparent abolishment of the relation that constructs the distinction between natural and human, or between human and animal.

Yet Clym's descent into this 'entangled bank', although it entails a loss of transcendent biological identity and a newfound proximity with nature, also conceals a movement of withdrawal. In becoming entangled with nature, he becomes disentangled from human relations: his 'daily life was of a curious microscopic sort,' writes Hardy, 'his whole world being limited to a circuit of a few feet from his person' (*Return*: 273). He becomes indifferent to his wife, who, dismayed at his 'social failure', becomes depressed and estranged from him — precipitating the fatal decision to resume her affair with Wildeve. Clym's notional loss of vision and identity as a response to the failure of his epistemological mastery seems in that sense narcissistic and to entail neither a connection with the infinite *epos* of germinal transformation not the egoic wound of abjection. Rather, it reproduces, covertly, the very type of split between human and nature it is thought to close: the imagined sense of community with non-human nature comes at the cost of departing from the *socius*, as if that itself were not natural.

We may see a similar dynamic *A Pair of Blue Eyes*, in what might be the most explicit reference in Hardy's fictional *œuvre* to the evolutionary sciences. The novel's famous 'cliff-hanger' scene stages a sensational series of events, during which the amateur geologist, Henry Knight, slips while on a coastal walk with Elfride and finds himself hanging from one of the steep promontories that line the Wessex coast. Knight, faced with almost certain death, is galvanised by a will to survive: 'He could only look sternly at Nature's treacherous attempt to put an end to him, and strive to thwart her' (*Blue Eyes*: 199). But his initial resolve is troubled as he notices, embedded in the cliff-face, a fossilised trilobite. The trilobite, an extinct, crab-like species, represents both evolutionary abjectness as well as an embodiment of the opposite of humanity's self-ascribed superiority, 'a low type of animal existence, for never in their vernal years had the plains indicated by those numberless slaty layers been traversed by an intelligence worthy of the name' (200). Faced with the ineluctable fact of his own imminent extinction, Knight's mind is borne imaginatively through the ancient past:

> Time closed up like a fan before him. He saw himself at one extremity of the years, face to face with the beginning and all the intermediate centuries

> simultaneously. Fierce men, clothed in the hides of beasts, and carrying, for defence and attack, huge clubs and pointed spears, rose from the rock [...]. Behind them stood an earlier band. No man was there. Huge elephantine forms, the mastodon, the hippopotamus, the tapir, antelopes of monstrous size, the megatherium, and the mylodon — all, for the moment, in juxtaposition. Further back, and overlapped by these, were perched huge-billed birds and swinish creatures as large as horses. Still more shadowy were the sinister crocodilian outlines — alligators and other uncouth shapes, culminating in the colossal lizard, the iguanodon. Folded behind were dragon forms and clouds of flying reptiles: still underneath were fishy beings of lower development; and so on, till the lifetime scenes of the fossil confronting him were a present and modern condition of things. (*Blue Eyes*: 200–01)

This remarkable scene has drawn extensive critical commentary, focusing on its genetic origins, as well as on its articulation of what might be called the evolutionary sublime.[40] For Carl Weber and Arthur McDowall it confirms Hardy's pessimism, by evoking the 'torture' of humanity's temporal insignificance, the 'personalised loneliness of nature'.[41] David Cecil notes its 'awe-inspiring vision[s] of infinite spaces and mysterious, irresistible forces'.[42] Beer too views Knight's 'companionship' with the trilobite as reflecting Hardy's concern with finding an appropriate scale for human life in an evolutionary ontology.[43] If Clym's encounter with nature's 'infinite space' is made possible by the relinquishment of sight and himself becoming invisible, we may focus by comparison on the way in which Knight's confrontation with the evolutionary sublime is premised on an extremely proximate form of seeing and on himself being seen in turn. The trilobite, Hardy writes, was 'a creature with eyes', which 'dead and turned to stone, were even now regarding him' back (200). While Clym's blindness allows him both to become entangled with and to disavow the natural world, for Knight it is an unwelcome form of seeing and being seen that precipitates an encounter with his own evolutionary origins. The past is inhabited in Knight's imagination, by 'monstrous', 'fishy', 'uncouth', and 'shadowy' forms, which repulse him, perhaps as representatives of an animality he would otherwise seek to ignore.

Can Knight's confrontation with the trilobite — an indexical signifier of evolutionary extinction — be understood as an experience of wounding self-abnegation in the face of evolutionary finitude? Certainly, he succeeds in facing his evolutionary origins and animality with 'open eyes' in a way that Clym does not. However, this encounter is a temporary incursion into Knight's ego. While he does briefly seem to fantasise about his own death, as foam-tipped waves appear to him as a 'white border to a black sea — his funeral pall and its edging' (*Blue Eyes*: 202), his morbid

40 For discussions of this passage's genesis, see: John Halperin, 'Leslie Stephen, Thomas Hardy, and "A Pair of Blue Eyes"', *Modern Language Review*, 75.4 (1980), pp. 738–45, doi:10.2307/3726582; J. O. Bailey, 'Hardy's "Imbedded Fossil"', *Studies in Philology*, 42.3 (1945), pp. 663–74; William J. Dawson, 'Thomas Hardy', in *The Makers of English Fiction* (Fleming H. Revell, 1905), pp. 213–40.

41 Arthur Sydney McDowall, *Thomas Hardy: A Critical Study* (Faber & Faber, 1931), p. 97; Weber, *Hardy of Wessex*, p. 36.

42 David Cecil, *Hardy the Novelist* (Constable, 1943), p. 73.

43 Beer, *Darwin's Plots*, p. 223.

reverie is cut short, as he is rescued from certain (symbolic) death by Elfride. The mode of his rescue is important too: Elfride disrobes, using her dress as a rope with which she drags him back to safety. Knight's survival is aided by Elfride's sexuality, a symbol of her desire for Knight's dominance and active masculinity. Whatever psychic disequilibrium may have been achieved by an encounter with deep time and extinction is cut short so that Knight's sexual predation can resume. In a sense, Knight's encounter with evolutionary death dramatises in a direct way the process of resistance to a Darwinian decentring of humanity where Clym's resistance is obscured. Knight, unlike Clym, seems to acknowledge the terror of the evolutionary sublime: the existential threat of evolutionary materiality for him justifiably precipitates a flight into a fantasmic humanity. For Knight, extinction repulses him, calls on him to abject himself as a subject wholly separate from the natural world. For Clym, natural entanglement does not cause horror: instead, Clym can disavow his human context, mis-recognising in nature a simplistically benevolent sphere of existence, thereby reproducing nature as a fetish object.

As in Zola's novels, tragic narrative causality and a failure of liberation can be linked to the structuring pessimism of Hardy's putative Darwinian worldview. If sexual selection seems to offer one kind of specifically gendered agency, the 'power of choice', this prospect is deflated in these two novels as Eustacia's and Elfride's attempts to transcend evolutionary determination are ultimately punished by destruction. Their transgressive desire, as well as Elfride's implicit biological feebleness, result in deaths that are figured as the predetermined outcome of indifferent forces. These novels also thematise sight, ways of observing, seeing, and not seeing, which promise to problematise the orthodox subject/object relation and thereby the distinction between human and animal (or its evolutionary origins). Yet even these efforts, by men rather than women, seem only to illuminate the way in which a desire to transcend evolutionary existence can result, or is haunted by, a disavowal of the evolutionary predicament. The imagined position of the all-seeing observer produces only further blindness. Communication with the animal under a veil of darkness invokes the same structure, albeit through a fantasy of self-obliteration in nature's embrace.

Sexual Selection, Excess, and Creative Possibility

Hardy's dramatisation of relations between the sexes and the problematics of vision seem to deny his characters the possibility of a liberation from the dismal meaninglessness and tragic determination of post-Darwinian life. If so, this outstrips even the pessimism of Zola's 'putrid' Naturalist fiction. For while Zola's revolutionary figures are destroyed or are themselves the agents of destruction, the destruction of the transgressive female figures in Hardy's fiction seems to be the product of comparatively narcissistic set of ambitions regarding not revolution or even reform, but personal love relations and desire. However, such a view dismisses the specific kind of agency instantiated by the women in these two novels. Elfride's and Eustacia's notionally personal desires may compare unfavourably in traditionally political terms with the male protagonists of Zola's novels, Claude, Florent, and Étienne, who each have grand ambitions for revolutionary transformation. Yet, it is noticeable that in Hardy's works, Clym's and Knight's attempts to evade the evolutionary real also offer a critique of that more conscious (and grandiose) desire, in specifically gendered terms. Noticing the way these novels, like Zola's, seem to gender messianic and egocentric conceptions of revolution in our relations with nature or animality, it might also be possible to read Hardy's gendered depictions of love relations as critical dramatisations of the way in which social forces seem to mimic or reinforce biological ones to constrain female agency. For Richard Kaye, Eustacia's vacillation in choosing either Wildeve or Clym suggests that the 'flirtatious female may, if she chooses, interrupt evolutionary "progress" as the novelist explores the forms of disaster that proceed from the overdetermined logic of sexual selection'.[44] Echoing Lawrence's view of the way in which stultifying Victorian social structures respond to 'aristocratic' rebels in Hardy's fiction, Kaye's argument is that such an effort to 'interrupt' evolutionary mechanism is necessarily met with destruction in the narratives, not simply as a way of reflecting (or indeed naturalising) the 'reality' of Victorian-Darwinian relations of the sexes, but to criticise that reality.

Another example of the way in which Victorian social mores and the appearance of evolutionary necessity respond to female choice can be found in the representation of the games of chess in *A Pair of Blue Eyes*. Two games of chess are represented in the novel: one, between Elfride and Smith, as this first courtship takes place, and later, between Elfride and Knight, as the possibility of their courtship comes into view. As such, the games can be read as a socially acceptable means of mediating sexual desire between Elfride and her suitors: a codified means of structuring advancement, retreat, advantage, and submission.[45] It also functions, in Darwinian terms, as a proxy competition between males, Smith and Knight, whose respective performances we are implicitly invited to compare. In the novel's first game of chess Elfride, for whom chess 'is [her] favourite scientific game', easily beats Smith, whose inexperience and comparatively lower social standing is reflected not only in his lack of skill, but also in the 'impression of indescribable oddness in Stephen's touch' (*Blue Eyes*: 156; 47). Knight, by contrast, easily defeats Elfride, whose mistakes in the game Knight capitalises upon in pitiless fashion. Smith too, by proxy, is beaten by Knight's superior skill and experience and these games map the power differentials and hierarchies between each competitor. In Darwinian terms, males compete for females, the more powerful, impressive, or skilful of them succeeds in vanquishing his competitor, as the female, in turn, gains access to a more suitable reproductive mate.

44 Richard A. Kaye, *The Flirt's Tragedy: Desire without End in Victorian and Edwardian Fiction* (University of Virginia Press, 2002), p. 257.

45 Mary Rimmer's reading of these games of chess points out that the meta-narrative of the chess game in particular is a gendered, rule-bound ritual that establishes and reproduces historically determined, patriarchal power relations: Mary Rimmer, 'Club Laws: Chess and the Construction of Gender in "A Pair of Blue Eyes"', in *The Sense of Sex: Feminist Perspectives on Thomas Hardy*, ed. by Margaret Higonnet (University of Illinois Press, 1993), pp. 203–19.

If these games, like the framed narrative or the painting in Zola, are to be read in relation to the wider narrative apparatus of the novel that contains them, they seem to echo the novel's rigid narrative determinism and troubling sexual politics. For if, as Darwin suggests in *On the Origin of Species*, the laws of sexual selection are less rigorous than those of natural selection, this principle seems only to apply to the male of the species in *A Pair of Blue Eyes*. Smith's defeat produces disappointment and shame; Elfride's loss to Knight is terminal. Her loss to Knight foregrounds his ultimate rejection of Elfride (on the grounds that she was previously involved with Smith) and, in so doing, diminishes Elfride's sexual value as a mate (in a sexual marketplace dominated by men), leaving her only the option of a loveless but necessary marriage with Lord Luxellian (and a pregnancy, it is assumed, that leads to her death). In these two competitions, the dynamics of sexual selection are figured in instrumental and hierarchical terms: as a game whose logic mirrors the zero-sum antagonism of social-Darwinian natural selection and not the aesthetic rituals associated with courtship and eroticism. As in Zola's Second Empire France, social and sexual taxonomies in Hardy's rural Wessex are taken for essential traits. By embedding the Darwinian narrative of sexual selection within clearly demarcated social and sexual hierarchies, Hardy's novel seems to suggest that social relations are driven by evolutionary necessity. Indeed, even Elfride's death, which seems narratively contingent on a characteristically Hardyean series of misfortunes and chance encounters, is reframed at the end of the novel as a 'miscarriage', implying its biological, and not its social, causality. However, might it not be possible to invert this reading as the imbrication of biological and social sexual hierarchies in the novel? If, as I have suggested, sexual selection offers a putative form of biologically grounded feminine agency in the form of choice, Hardy's novel draws attention to the way that choice is constrained not by natural laws but because the social context of its unfolding constrains any radical or transgressive potential. Elfride's fate is predetermined by the social expectations of her gender while the qualified agency accorded to the phallic sexual males is confirmed by pre-existing social hierarchies, according to which the most powerful male figure, Lord Luxellian, 'wins'.

It follows, therefore, that if Hardy's work is implicitly critical of the naturalisation of biological theory by social relations, we should also attend to the way female agency in these works — as reflected in representations of *sexual* relations and what Kaye calls 'flirtation' — might be understood in affirmative terms. In that light, let us return to re-read Eustacia's role as voyeur in *The Return of the Native*. Her part in the mummers' play provides her with a form of social and epistemological positioning usually reserved for the scientific observer. Behind the mask of her costume, she assumes the capacity to both take part in and remain separate from a constellation of actors and her audience, simultaneously performing and observing, being watched and watching. Her concealment performs an ambiguously gendered form of sexual desire. For in taking on a disguise, she not only plays with the distinction of seeing and being seen but also with the fixity of gender binaries and their relation to biological sex. In this novel's mummers' play, the traditional folk play of St George and the Turkish Knight, Eustacia takes on the role of St George's adversary, her costume concealing her biological sex: 'revealing herself to be changed in sex, brilliant in colours, and armed from top to toe' (*Return*: 163).[46] This implicitly sexual relation transforms Eustacia and is itself transformed. Eustacia's anonymity allows her to act as voyeur, casting her in the role of a phallic aggressor, an avatar of male power (albeit one whose identity is implicitly feminised and racialised as the 'Oriental' other), and as a pugnacious and ostentatiously decorated male competitor. Eustacia's gender performance inverts the Darwinian image of the sexual competition, which is emblematised by the avian mock battles in *The Descent of Man*:

> When we treat of sexual selection, we shall meet with innumerable instances of this form of transmission, — as in the case of the spurs, plumes, and brilliant colours, acquired by male birds for battle or ornament, and transferred to the females in an imperfect or rudimentary condition. (*Descent*: I, 208)

Like Darwin's birds, Eustacia's dramatic performance of a battle with the Valiant Knight of Christianity is part of a decorative arms race for sexual attention. The narrator outlines the conventions of this folk ritual:

> It might be that Joe, who fought on the side of Christendom, had a sweetheart, and that Jim, who fought on the side of the Moslem, had one likewise. During the making of the costumes it would come to the knowledge of Joe's sweetheart that Jim's was putting brilliant silk scallops at the bottom of her lover's surcoat, in addition to the ribbons of the visor, the bars of which, being invariably formed of coloured strips about half an inch wide hanging before the face, were mostly of that material. Joe's sweetheart straight-way placed brilliant silk on the scallops of the hem in question, and, going a little further, added ribbon tufts to the shoulder pieces. Jim's, not to be outdone, would affix bows and rosettes everywhere. (*Return*: 158)

This passage makes clear that the play, as well as thematising competition between men, is also part of a meta-theatrical courtship structure in which the spectators are involved in a decorative one-upmanship, in which male costume is a vector for female competition in relation to social and aesthetic status. Eustacia — no longer the 'sweetheart' but in the position of Joe or Jim — usurps the role of the 'man' or rather reveals the role of the male to be performative, bound up with the aesthetics and spectacle of war. Her position, therefore, is highly complex: there are multiple, competing, and self-concealing performances of gender folded into this theatrical performance and its para-theatrical preparations, to which Eustacia adds her own gender-ambiguous subversion of the role. If we are to understand Hardy's courtship narratives as reflections of a narrative of sexual selection, its basic relational structure — sexual desire as enacted through forms of competition that involve relations of attraction and desire — also provides the frame for a subversion of conventional Victorian conceptions of gender and sexuality to which it is linked.

46 Ruth Firor offers a detailed study of folk practice in Hardy, including that of the mummers' Christmas play in *The Return of the Native* which, she writes, 'is interesting for the glimpse it affords us of folk-drama presented in a strictly traditional manner': Ruth Firor, *Folkways in Thomas Hardy* (University of Pennsylvania Press, 1931), p. 152.

In so doing, this scene troubles a reading of sexual selection in Hardy according to which his work (and, by extension, that of nineteenth century realism *tout court*) uncritically reflects and implicitly reproduces biological accounts of sexuality and their patriarchal subtexts. In this scene, Hardy's vision of competitive sexual performance can be understood both to engage with a Darwinian conception of sexual imperative while also, implicitly, complicating the corporeal and sexual essentialism that Darwinism is assumed to underwrite. For Grosz, this juxtaposition of gender trouble and evolution is not necessarily contradictory; Darwin's work 'provides feminist and cultural theory with a way of reconceptualising the relations between the natural and the social, between the biological and the cultural, outside the dichotomous structure in which these terms are currently enmeshed'.[47] Criticisms of his anti-feminism or patriarchy, she suggests, may be historically justified but share with their object of critique the assumption that 'nature' is necessarily a obstructive concept to feminist liberation, reinscribing the implicit anthropomorphism of some feminist concepts of liberation.[48] Instead, for Grosz, like Deleuze, Darwin offers us a resource with which to develop an image of nature that is in flux and in dynamic interrelation with culture. Specifically, in Grosz's work, sexuality represents a principle of excess in the natural world and sexual selection functions as a supplementary dynamic that disturbs the logic of natural selection, by adding to and complicating its ruthless logic of elimination. Evolution should not be understood as a calculable and regular dialectic of death and growth, she argues, but as an indeterminate process of the unfolding of life, in which reproduction (understood in its broadest sense) gives rise to a host of unpredictable, emergent cultural as well as natural forms, whose interaction deflects any purely instrumental, deterministic conception of evolution.

Underpinning this reading of Darwin is George Bataille's distinction between 'restricted economy' which functions according to 'determinable rules and procedures' and 'general economy', characterised by excessive forms of gift-giving and waste. Grosz draws an analogy between the dynamics of natural selection and calculable reasoning of Bataille's 'restricted economy', both of which function according to a determinable logic of protocols and outcomes. In contrast, she argues, sexual selection functions 'according to a general economy, without order, without striations or organization'.[49] For Bataille, the waste and gift-giving of the general economy are exemplified by art and by non-procreative (or erotic) sexuality, which reveal the restrictive and nominally rational economics of exchange and growth to be inadequate to 'cosmic' or complex systems of human

47 Grosz, *The Nick of Time*, p. 91.

48 Elizabeth Grosz, 'Darwin and Feminism: Preliminary Investigations for a Possible Alliance', *Australian Feminist Studies*, 14.29 (1999), pp. 31–45 (p. 31), doi:10.1080/08164649993317; For feminist critiques of Darwinian sexual selection see: Ruth Hubbard, *The Politics of Women's Biology* (Rutgers University Press, 1990); Sue Rosser, *Biology and Feminism: A Dynamic Interaction* (Twayne, 1992), p. 57; Patricia Gowaty, 'Introduction: Darwinian Feminists and Feminist Evolutionists', in *Feminism and Evolutionary Biology: Boundaries, Intersections, and Frontiers*, ed. by Patricia Gowaty (Chapman and Hall, 1997), pp. 1–18 (p. 5); Carol Colatrella, 'Feminism at War: Sexual Selection, Darwinism, and Fin-de Siècle Fiction', in *After Darwin*, ed. by Griffiths and Kreisel, pp. 108–20.

49 Grosz, *Becoming Undone*, p. 131.

relation. For Grosz, then, if natural selection can be understood as a form of 'restricted economy', sexual selection exemplifies Bataille's conception of 'general economy' in its irreducibility to Malthusian cost-benefit calculations. According to Bataille, the restrictive economic method of arbitrating expenditure based on a perceived, immediate benefit in return, in a general economic sense, fails to take into account the ineffable and unpredictable benefits of surrendering wealth with no expectation of profit in return.[50] Grosz conceives of sexual selection as a form of sacrificial eroticism, which is to say, like Bataille in his work on the sacrificial nature of sexual desire, she differentiates the 'psychological quest' of exuberant sexuality and pleasure from its assumed telos of reproduction. '[E]roticism,' Bataille writes, 'is the sexual activity of man to the extent that it differs from the sexual activity of animals'. By animal sexuality, Bataille here means the sexuality of the restricted economy, conducted, unreflexively, with the intended outcome of reproduction and, thus, although 'human sexual activity is not necessarily erotic [...] erotic it is whenever it is not rudimentary and purely animal'.[51] Grosz draws on and inverts Bataille by suggesting, via Darwinian sexual selection, that what is specifically human is merely a fantasy of sexuality as a solely procreative cost-benefit relation (both for animals and for humans), when in fact all sexuality is by necessity excessive, since not only humans but also non-human species engage in a host of sexual rituals, performances, and aesthetic relations, which cannot be reduced to the *realpolitik* of natural selection.

For Grosz, therefore, it is vital that philosophy accepts that sexual excess is not the purview of 'culture' (or Freudian 'civilisation'), or a metonym for 'humanity', but is seen as an integral aspect of the 'natural'. Her focus is on how erotic desire and sexual play can be understood as arising from and integral to the necessity of evolutionary reproduction (the animal drive) while also engendering various forms of desire or relation that are enjoyed for their own sake, generate unpredictable outcomes, and/or broaden out the field of sexual signification. 'The laws of sexual selection,' she argues, 'are the principles of aesthetics, not the strategies of game theory; the functioning of appeal rather than the operations of rational agents who act according to their self-interests; the order of taste rather than [...] calculation'.[52] This is a development of Darwin's thinking in *The Descent of Man* where, detailing the sexual rituals of the bowerbird, he describes the way sexuality is premised on aesthetic preference:

> Why certain bright colours and certain sounds should excite pleasure, when in harmony, cannot, I presume, be explained any more than why certain flavours

50 'Changing from the perspectives of *restrictive* economy to those of *general* economy accomplishes a Copernican transformation: a reversal of thinking — and of ethics. If a part of wealth (subject to a rough estimate) is doomed to destruction or at least to unproductive use without any possible profit, it is logical, even *inescapable,* to surrender commodities without return. Henceforth, leaving aside pure and simple dissipation, analogous to the construction of the Pyramids, the possibility of pursuing growth is itself subordinated to giving': Georges Bataille, *The Accursed Share: An Essay on General Economy. Volume 1: Consumption*, trans. by Robert Hurley (Zone Books, 1988), p. 25.

51 Georges Bataille, *Eroticism*, trans. by Mary Dalwood (Penguin Classics, 2012), p. 29.

52 Grosz, *Becoming Undone*, p. 131.

> and scents are agreeable; but assuredly the same colours and the same sounds are admired by us and by many of the lower animals. (*Descent*: I, 64)

Darwin draws attention not only to the aesthetic sense and experience of pleasure shared by both animals and humans, but also to its origin in seemingly subjective and obscure reasoning. While the 'choices' made by individuals in an economic or competitive biological field can be understood as rational (in the sense that one is required to maximise one's chances of survival and profit), here Darwin is unable to provide similar reasoning as to the nature of sexual and aesthetic preferences (both in human and animals).

In this light, we can read Eustacia's and Elfride's failures to succeed as evolutionary agents affirmatively. Both women die and fail to reproduce not primarily because, like Zola's revolutionaries, they are biologically determined, but because they follow contingent and inexplicable desires, refusing to attenuate these either to the restrictive logic of natural selection or to the equally punitive sexual norms of Victorian society. Elfride and Eustacia are Bataillean sacrificial figures, women who affirm the possibility and reconfirm the risk of exceeding the limited logic of biological instrumentality. For Grosz, this agency can be found in the excess of bodily modifications and intensifications of colour which Darwin's conception of sexual selection postulates in exorbitant morphologies, through song, through dance, but also through territorial practices like nest-making and self-adornment. The human body becomes a site and generator of biocultural transformation and the supposedly superficial (read: feminine) adornments of hairstyle or jewellery, are, in the arena of sexual selection, integral elements of human morphology whose very existence complicates the idea of bodily purity and evolutionary necessity. Judith Butler's theorisation of bodily modification and its relation to biology offers a useful touchpoint in developing the radical nature of the thesis Grosz finds in Darwin, and which is articulated in Hardy's narratives of sexual selection.

> The construction of stable bodily contours relies upon fixed sites of corporeal permeability and impermeability. Those sexual practices in both homosexual and heterosexual contexts that open surfaces and orifices to erotic signification or close down others effectively reinscribe the boundaries of the body along new cultural lines.[53]

Understood in these terms, we can interpret the performativity of Eustacia's gender play as a way of destabilising the notion of corporeal stability, by instantiating forms of prosthetic or non-natural erotic signification. These forms, expressed through costume and self-adornment, trouble biologically essentialist and hetero-normative readings of Darwinian sexual selection by both acting out the narrative of sexual competition the Darwinian model provides while, to reuse Butler's words, 'reinscrib[ing] the boundaries of the [post-Darwinian] body along new cultural lines'.

More expansively still, Grosz argues that this way of understanding sexual selection and erotic desire indexes the origins of art: 'the creation of music and

53 Judith Butler, *Gender Trouble* (Routledge, 2002), p. 169.

art, visual display, and the joy of immersion in sonorous or visual qualities, are a primordial resource of sexual selection'; and that art is merely 'the formal structuring or framing of [...] intensified bodily organs and processes which stimulate the receptive organs of observers and coparticipants'.[54] In the context of reading Hardy, this analysis complements his works' implicit critique of Victorian patriarchal and biologically determined restrictions with a conception of Darwinian evolution. Rather than focus only on how narrative endings in Hardy's fiction reconfirm notionally Darwinian and pessimistic sexual fatalism, thereby limiting our reading of sexual selection to the restrictive economy, we may instead focus on what relations between the sexes create — cultural, aesthetic, and affective forms which add to a general economy and that themselves transform the social and the biological.

On this count, while we can see both Elfride and Eustacia as revolutionary and sacrificial figures, *A Pair of Blue Eyes* and *The Return of the Native* may offer an instructive contrast in the way they each dramatise the aesthetics of Darwinian sexual selection. Elfride's relations with her prospective lovers, as exemplified by the games of chess, seem to remain trapped within an instrumental, rules-bound, form of relationship according to which the telos — win/loss, marriage/failure — retroactively determines the value of the ritual. As such, her fate is determined not only by the zero-sum rules of competition, but also by the ideological coordinates of that competition, according to which women are passive objects for male competition, with reproduction and heterosexual marriage as its aim. This is suggestively thematised in *A Pair of Blue Eyes* by Elfride's earrings and their significance in shaping the tragic direction of her narrative. After Smith and Elfride's mutual parting, later in the novel Knight tests her suitability as a mate by offering her the choice between 'the very prettiest earrings in Bond Street windows' versus a 'well-chosen library of the best music'. Elfride (despite Knight's masculine dismissal of the former as 'jimcrack jewellery') chooses the earrings, having 'lost one of [her] prettiest pair last year', when walking with Smith at the moment they share a first kiss (*Blue Eyes*: 170). The dynamics of sexual selection, therefore, are structured around bodily enhancement. However, where earrings seem to have been used by Elfride to attract Smith earlier in the novel (that is in an inversion of the plot of sexual selection regarding male bodily enhancement), her loss of one of these earrings functions as a narrative prolepsis for Knight's confirmation of her notional feminine vanity, as does his later discovery of her relationship with Smith (which diminishes her sexual value for him), when he happens upon the lost earring. Bodily enhancement does not function here to enhance or multiply Elfride's sexual or agential possibilities, but as an instrument of patriarchal control and possessiveness.

By contrast, in *The Return of the Native*, while the mummers' play, like the game of chess, can be read as a socially acceptable way of ritualising and therefore structuring sexual desire, it fails to essentialise sexually heterodox relations but, perhaps despite itself, creates a space in which these can be challenged. Eustacia transmutes from

54 Grosz, *Becoming Undone*, p. 135.

female to male in a performance decoupled from her sex. In her role as the Turkish Knight, she gives a virtuosic, sexually ambivalent performance as a phallic violent male soldier, enhancing his/her appearance both for martial and sexual purposes, underneath which lies, indeed which makes possible, female desire and the exercise of sexual choice. Self-adornment and performance, therefore, transform the terms of engagement, refusing corporeal or biological essentialism and allowing Eustacia to observe her object of desire, and casting Clym, the notional male (at least in terms of biological sex), in the 'feminine' position as a passive spectator, unaware of his own objectification. If it can be argued that Eustacia is required to become male to enact her desire, what is important is nevertheless the plasticity of gender performance and its decoupling from biological sex. The play itself, the battle between the Turkish Knight and St George, thematises violent antagonism, like chess, but the absence of formal rules, or a specific telos of the play, its performative openness, paves the way for an enactment of sexual selection that subverts the orthodox sexual dynamics of Victorian, scientific ideology.

So, the narrative of Darwinian sexual selection is subverted by the character of Eustacia, through gender performance, androgyny, and the transgression of heterosexual hierarchy. How might this reflect upon the Darwinian concept of sexual difference as such? The answer to this question, for Grosz, lies in the foundational distinction made between the human and the animal. In much philosophical and theoretical literature, Grosz asserts, the animal is understood to be incapable of rationality, of reflection, and, therefore, of ethics, and this distinction is extended metonymically to women, children, slaves, and others, as 'the alignment of the most abjected others with animals is almost ubiquitous'.[55] Leaving intact this primordial distinction allows us, humans, to constitute ourselves as rational, reflexive, and ethical — a distinction, Grosz argues, from which all further oppressive distinctions of identity flow. As such, Grosz's work seeks not to deconstruct representation only to reconstruct a proliferation of new essential identities, but to explore 'difference as the generative force of the world, the force that enacts materiality (and not just its representation), the movement of difference that marks the very energies of existence before and beyond any lived or imputed identity'.[56] Like Deleuze, therefore, Grosz sees Darwin's work not only as an intervention in humanity's biological continuity with and derivation from the animal, but also as a theoretical resource in thinking a form of ontological difference which precedes concretised differences.

55 Ibid., p. 12.

56 It is on this question of signification and being that Butler's and Grosz's approaches to sexual difference seem to diverge, a theoretical and methodological distinction that has caused a significant amount of controversy, particularly as regards its implications for theorising trans- or non-binary identities. For some, Grosz (and Luce Irigaray, to whom Grosz owes this ontological turn), by theorising *sexual difference* as an ontological principle of creation, reassert the primacy of a fundamentally essentialist conception of sexuality (and 'nature') over the performative conception of gender. I postulate that these seemingly mutually-opposed conceptions of gender seem to share a commitment to creativity and change as a condition of a nature in dynamic conversation with culture. See Pheng Cheah and others, 'The Future of Sexual Difference: An Interview with Judith Butler and Drucilla Cornell', *Diacritics*, 28.1 (1998), pp. 19–42; Grosz, *Becoming Undone*, p. 91.

This pre-representational conception of difference emerges in part from Deleuze and in part from the work of Luce Irigaray, from whose work Grosz develops a consciously feminist iteration of Darwinian notion of becoming.[57] If for Darwin, sexual difference is the catalyst for biological reproduction (and change), for Irigaray, 'becoming' is itself sexed:

> The natural is at least two: male and female. All the speculation about overcoming the natural in the universal forgets that nature is not *one*. In order to go beyond — assuming this is necessary — we should make reality the point of departure; it is *two* [...]. The Universal has been thought as one, thought on the basis of *one*. But this *one* does not exist.
>
> If this *one* does not exist, limit is therefore inscribed in nature itself. Before the question of the need to surpass nature arises, it has to be made apparent that it is *two*. This *two* inscribes finitude in the natural itself. No one nature can claim to correspond to the whole of the natural. There is no 'Nature' as a singular entity.[58]

For Grosz, Irigaray's theory of a non-representational form of pre-physical sexual difference lends further philosophical dynamism to Darwin's notion of sexual selection and opportunity to engage with rather than overcome 'the natural'. Difference as a sexually bifurcated ontological condition is echoed in but not the same as the sexuate manner through which the biological world reproduces itself, and the structural dynamics of which are described in Darwin's work on sexual selection. In this respect, Irigaray distinguishes between 'sexual difference' ('la différence sexuelle') as a universal, ontological precondition for the elaboration of difference and 'sex difference' ('la différence des sexes'), as the empirical morphological difference between sexually dimorphic organisms.[59] If (sexed) ontological difference can be understood as what Grosz calls the 'ineliminable' but decentred condition through which social and natural relations arise and proliferate, sexual selection for the purposes of reproduction, by contrast, is limited to those relations and encounters between entities characterised by binary sexuate morphologies. Yet, where these two nominally separate realms come into contact, Grosz argues, is precisely through the excess that sexual selection generates — self-adornment, aesthetic production, desire's affective charge — in which the very barrier between nature and culture becomes overturned.[60] Sexual selection, therefore, though limited to the dynamics and relations that structure biological reproduction,

57 Elizabeth Grosz, 'The Nature of Sexual Difference: Irigaray and Darwin', in *Sexual Difference between Psychoanalysis and Vitalism*, ed. by Arun Saldanha and Hoon Song (Routledge, 2014), pp. 69–93.

58 Luce Irigaray, *I Love to You: Sketch of A Possible Felicity in History*, trans. by Alison Martin (Routledge, 1996), p. 35.

59 Alison Stone, *Luce Irigaray and the Philosophy of Sexual Difference* (Cambridge University Press, 2006), p. 94.

60 Darwin's theory of sexual selection, Grosz argues, 'can be understood as an analysis of the proliferation of nothing but differences: differences without any hierarchical order, without fixed identities or biological archetypes; differences generated for their own sake and evaluated only through social and natural contingency differences without norm, without inherent value': Grosz, *Becoming Undone*, p. 167.

troubles the difference between nature and culture, according to which evolutionary reproduction involves bodies becoming cultural as culture becomes natural. It also offers, in its echo of the 'nature which is not one', a means through which an ethical or feminist effort to surpass or transcend the natural, whether understood as sexed reproduction or ontological difference, can be replaced with an embrace of the way these separate spheres of creation are intertwined.

How can such a philosophical intervention in our reading of Darwin inform our understanding of the way in which the narratives of sexual selection subvert historical (and existing) cultural and scientific sexual and gender normativity? As critics have pointed out, Irigaray's and Grosz's conceptualisations of sexual difference as an ontological condition risks re-naturalising a relational binary which is not only a fantasy in human social relations but in non-human relations too.[61] Myra Hird identifies the kind of anti-representational feminism found in these thinkers as a new 'ontology of heterosexuality' and points out that Irigaray's imputed empiricism fails on its own count, since 'most organisms' on earth 'do not reproduce through sexual reproduction'.[62] As such, to insist on sexuate difference as a molecular, ontological, metaphysical, or even divine feature of life reproduces the same kind of humanism that the work of Grosz and Irigaray attempts to repudiate. Drawing on the groundbreaking work of Lynn Margulis, Hird demonstrates that the bifurcated sexual difference we see in human sexual relations (and in many other species) is itself evolutionarily (and therefore deep-historically) contingent, one of many possible solutions to the demand for organic life to reproduce itself (and therefore to evolve). Biological reproduction, Hird says, may occur in many different ways ('cosmic irradiation, virus and symbiont acquisition, or exposure to ambient chemicals') which trouble not only a Victorian conception of sexuality, but the putatively radical re-conception of Darwinian sexual selection offered by Grosz via Irigaray.[63]

Eustacia's gender performance can be understood as a gesture of representation, a play with the signifiers of well-established, folkloric gender roles, which puts into question the simplistic evolutionary scripts of heteronormative sexual selection in Hardy's work. However, it is precisely through this vector of representation (or more appropriately, signification, a term that does not imply a fixed referent) in Eustacia's performance that the radical potential of Grosz's re-theorisation of Darwin might be elaborated. The constellations and relations that sexuate reproduction and sexual selection seem to construct — heterosexual love and homosocial competition — are undermined here by the forms of signification, aesthetic play, and gender performativity that occur within those structures. The excess generated by sexuality need not be ontological; for Eustacia, the excess (or waste) in her sexual performance is a function of signifying practices, which supplement (and therefore reveal as incomplete) the standard script of a socially or biologically determined

61 Abigail Bray, 'Not Woman Enough: Irigaray's Culture of Difference', *Feminist Theory*, 2.3 (2001), pp. 311–27, doi:10.1177/14647001010200304; Schaefer, 'Darwin's Orchids'.

62 Myra J. Hird, 'Digesting Difference: Metabolism and the Question of Sexual Difference', *Configurations*, 20.3 (2012), pp. 213–37 (p. 224).

63 Ibid., p. 224.

structure. If Irigaray's and Grosz's work criticises attempts to 'overcome nature' by a retreat into representation, sexual selection (as Grosz's work, ironically, helps us to see) implies that signification — whether through colour, form, disguise, costume, as well as language — is itself natural. Yet, Grosz's work does help us to see that Darwinian sexual selection is more complicated than many contemporary and historical accounts of the so-called sexual marketplace. She invites us to read Darwin's elaboration of sexuality not as a theory of rational computation within a game theoretical framework, but as a counter-hierarchical, trans-individual, and dynamic set of relations.

Hardy's work stages the subjective and transindividual nature of sexual relations in *The Return of the Native*'s representation of another folk ritual: the 'gipsying' folk dance in which Eustacia takes part with her former (and future) lover, Wildeve. Like chess and the mummers' play, the village dance functions to codify sexual desire in forms that echo the narratives and hierarchies that feature in Darwinian sexual selection. The village fete is dominated by 'the grand central dance of fifteen or twenty couples, flanked by minor dances of inferior individuals whose gyrations were not always in strict keeping with the tune' (*Return*: 279). The young men, we are informed, 'wore blue and white rosettes, and with a flush on their faces footed it to the girls'; the women, in response, 'blushed deeper than the pink of their numerous ribbons' and wore curls, 'lovelocks', and braids with which to attract their suitors (279). The dance is, therefore, another paradigmatically Darwinian ritual, reminiscent of the 'young rustics at a fair', the spectacle of whose sexual competition Darwin compares to that of the courtship rituals of birds. However, at this point of the narrative, Eustacia's objective is not self-consciously sexual but psychically reparative. Her presence at the dance is framed as a response to the gloom caused by her husband's social alienation and involves a series of affectively ambiguous states: 'I'll be bitterly merry,' she asserts, 'and ironically gay, and I'll laugh in derision. And I'll begin by going to this dance on the green' (279). Unlike her husband, then, whose melancholy commune with the natural conceals a withdrawal from the social, Eustacia, in response, is characterised not by a form of refusal but by a complex (or perverse) form of participation in a collective and creative act — a creative act that can be understood as both social and, after Darwin, biological. Eustacia, therefore, by participating (rather than withdrawing), instantiates an ambiguity of feeling, as against the binary states of terror and bliss that define Clym's and Knight's encounters with the evolutionary sublime.

It is here, then, in the everyday creativity of folk dances and courtship rituals, rather than in the conscious and unconscious attempts by phallic subjects and messianic revolutionaries, that evolutionary determination is halted or complicated. The ritual of sexual selection, which appears in the earlier work, *A Pair of Blue Eyes*, as a form of hetero-normative and reproductive realpolitik, appears in this novel as a space of affective and creative possibility:

> The enchantment of the dance surprised her. A clear line of difference divided like a tangible fence her experience within this maze of motion from her experience without it. Her beginning to dance had been like a change of

> atmosphere; outside, she had been steeped in arctic frigidity by comparison with the tropical sensations here. She had entered the dance from the troubled hours of her late life as one might enter a brilliant chamber after a night walk in the wood. (*Return*: 282)

The sense of liberation and enlightenment in the dance versus the constraint and gloom outside are articulated a series of symbolic binaries (hot/cold, motion/rigidity, light/dark) which are separated in a 'tangible' way by the experience of the folk dance. In this scene, therefore, we may read Eustacia as becoming participant in what Miller calls in Hardy the 'magically charged milieu' through which life instinct is experienced affirmatively: as she and Wildeve begin to dance, 'through the length of five-and-twenty couples they threaded their giddy way, and a new vitality entered her form' (*Return*: 283).[64]

Such vitality, of course, may be thought as delusional — or perhaps simply insignificant — as this momentary break or gap in biological and social determination is folded back into the narrative of Eustacia's death. However, to privilege the end of the narrative line, as opposed to what is created in its interruption, also privileges the presence of Eustacia (which the text achieves through its narrative focalisation), in what is understood to be a collective and not a merely individual act of creation. Through its setting with the community dance, Eustacia's doomed engagement with Wildeve is set in a wider narrative context as 'but one of the many of those impassioned but temporary embraces [...] destined to become perpetual' (*Return*: 281). The gipsying dance, therefore, is a way for a community of lovers to give expression to and partake in a ritual that facilitates erotic relation, creating forms of affective excess that cannot be reduced to the calculable mechanics of biological reproduction: 'A whole village-full sensuous emotion, scattered abroad all the year long, surged here in focus for an hour' (281). If as Beer points out, Hardy's narrative focus on the single generation (and its termination) represents a form of tragic humanism, this marginal reference to the communal experience of erotic relation, and the series of unknown narrative ends to which it refers, suggests a more capacious and narratively ambiguous focus on human fate. Deleuze's reading of Zola emphasises the supra-human force both of biology and revolutionary transformation, but the village in which Eustacia and Wildeve play out their dance represents a less cosmically pregnant, but still radical, horizon of possibility. If Eustacia, we know, is destined to be 'swallowed up' by biological and social fate, this does not abnegate the way the dance produces unknowable different possibilities for those other characters who participate, and which may also provoke unpredictably new kinds of feeling and desire that are not reducible to an instrumental conception of sexual selection. Hardy's work here might be seen as focusing on surplus or excess characters, figures that by being marginal to the narrative seem to evade its hereditary determination.[65] As such, the absence of endings of these 'merely' marginal characters indexes a social (and biological) world external to the one that Hardy's tragic heroines inhabit; this, in turn, leaves

64 Miller, *Thomas Hardy*, p. 144.

65 Emily Steinlight, 'Hardy's Unnecessary Lives: The Novel as Surplus', *Novel*, 47.2 (2014), pp. 224–41, doi:10.1215/00295132–2647167.

open the possibility — however unlikely the narrative might make it seem — that the erotic relations produced by the sexual can, in the words of Grosz, 'derail' the locomotive force of natural selection's determinist death drive. Hardy's novel by making a gesture at a social totality, shifts our focus to lives other than those of the novel's protagonists, the absence of whose narrative telos indexes an uncertainty that is integral to the rituals of selection.

Conclusion

Hardy's claim that his writing could be read as a form of 'evolutionary meliorism' suggests a discomfort with the attribution to his work of a biologically determined and pessimistic *Weltanschauung*. Yet, in 'Candour in English Fiction' he is more forthright in his defence of tragedy, specifically in opposition to the 'puerile inventions' and 'thirst for accuracy' which characterised, in his view, the quasi-biological theoretical foundations and encyclopaedic aesthetics of literary Naturalism.[66] Here, as in his essay on 'The Science of Fiction', Hardy seeks to reserve for literary art an autonomy from the requirements of scientific naturalism and scientific representation, and instead to articulate the conception of a dynamic relation between art and natural life. Hardy finds in Greek tragic theatre, in particular, an exemplar of literary autonomy and dynamism which, in his reading, did not attempt to enumerate the world merely accurately or exhaustively, but rather sought to uncover something in life that previously was invisible. 'They reflected life', Hardy says of the Greek tragedians, and in doing so 'revealed life, criticised life'.[67] If in Hardy's tragic fictional world fate is imprisoned by biology, it is precisely for this reason that his writing can be understood as a form not only of realist 'reflection' or intellectual 'revelation' but also of immanent critique. If his narrative cosmology seems irrevocably determinist in its treatment of those who seek to escape the bounds of social and biological fate, it is from within this position, and through the Darwinian narrative of sexual selection, that a critical view of this world is articulated.

Hardy's novels, as those by Zola, mount such a critique from within the Darwinian logic of life in several forms. The first, as it appears in this chapter, takes the form of tragic narrative as a form of biological fate, which I have argued reaffirms the tragic determinism of Darwinian narrative. Yet, we may also read this another way: as a mode of bearing witness to the brutality of human life as determined by the natural laws of Darwinian competition, in terms of survival, the failure of meliorism, and the delineation of tragic, biologically determined sexual relations. It is true that if Eustacia and Elfride are fated by the logic of Darwinian and Victorian patriarchy to be extinguished for being deviant, the effect of that might be to naturalise a particularly deterministic conception of post-Darwinian human life, and its gendered, sexualised valences. Yet, if Hardy's work can be considered 'realist' in the sense of reflecting 'the real', it does so not in offering a representational or realistic account of post-Darwinian life in Victorian England

66 Hardy, 'Candour in English Fiction', p. 97.
67 Ibid.

but, like that of Zola, through its insistence on tragedy and, in Hardy's work, tragic endings for women. Such an approach, in tandem with an understanding of Darwin's significance for Hardy, works to *de*-naturalise the Darwinian real, allows us to read Hardy's work as a dramatisation of a reality constructed by a Darwinian determinist concept of life.

This concern with realism and its capacity to bear witness to humanity's animality is developed and problematised in Hardy's thematisation of vision, in relation to the human subject's encounter with an evolutionary real. In these two novels, vision and observation can be understood as ways for the subject both to bear witness to, and in turn to transcend, their positions as post-Darwinian human animals. Those efforts, by Clym and by Knight, are revealed in these novels to be the functions of a fantasy according to which the human subject is either wholly swallowed up by or removed from evolutionary materiality (as indexed by the trilobite fossil or the flora and fauna of Clym's furze-cutting fields). In the character of Eustacia, however, we are provided with a more promising understanding of vision's relationship with the natural world, through her suggestively transgressive covert observation of Clym, a position that both enacts and complicates the agonistic positions afforded by the narrative of sexual selection. Here, the realist observer is a participant in the world she observes, alive to the possibility of being seen by her object as much as putting herself in the position of a subject. In her role as the Turkish Knight, moreover, Eustacia takes up a subversive position in relation to the cultural and biological scripts that, at the level of tragic narrative fatalism, are understood to determine her life. As such, Hardy's work can be read as going beyond a critique of a notional Darwinian narrative of biological determination and a revelation of evolutionary violence (which defines the earlier novel) by articulating through the figure of Eustacia a conception of evolutionary sexuality that resists the rigid normativity prescribed by Victorian morality.

And it is through the character of Eustacia and her engagement with ritual forms of erotic relation that we can glimpse a vision of a theoretically radical conception of sexual selection. Eustacia's subversion of the standard narrative of sexual selection can be understood through the work of Grosz to represent a more faithfully radical interpretation of Darwin's conception of sexuality, according to which sexual selection generates a cultural, affective, and creative excess that transforms the biological field. Whether this is understood as a function of a fundamentally sexuate ontology, as per Irigaray, or as a way for re-reading Darwin's work on sexuality in relation to later developments on the performative nature of gender, Eustacia enacts a form of creative erotic relation conceived not as a form of withdrawal or transcendence either from nature or culture, but as a form of participation in which the cultural and the natural are intertwined. This does not exempt Eustacia from her fate as a sacrificial victim of evolutionary determination but underlines the collective nature of the folk dance, in which Hardy's work gestures at a wider set of narrative possibilities beyond those of Eustacia and Wildeve. As such, the collective nature of evolutionary eroticism in Hardy's work and its integration in a scene of everyday folk creativity invite us to reconsider the possibilities afforded by trans-individual relations within a Darwinian conception of anti-determinist possibility.

CHAPTER 3

❖

'Dreaming of Islands': Three Darwinian Utopias[1]

> Dreaming of islands — whether with joy or in fear, it doesn't matter — is dreaming of pulling away, of being already separate, far from any continent, of being lost and alone — or it is dreaming of starting from scratch, recreating, beginning anew. — GILLES DELEUZE[2]

> The very close relation of the distinct species which inhabit the islets of the same archipelago, — and especially the striking relation of the inhabitants of each whole archipelago or island to those of the nearest main-land, — are, I think, utterly inexplicable on the ordinary view of the independent creation of each species, but are explicable on the view of colonisation from the nearest or readiest source, together with the subsequent modification and better adaptation of the colonists to their new homes. — CHARLES DARWIN (*Origin*: 406)

Introduction: Desert Islands

In an early, posthumously-published essay, 'Desert Islands', Gilles Deleuze elaborates a critique of the yearning in Western thought for transcendence and its sublimation as the meta-poetic figure of the desert island. The space of the desert island in Daniel Defoe's *Robinson Crusoe*, Deleuze points out, represents a geo-philosophical myth of bourgeois individualism and *ex nihilo* creativity while unveiling that myth's fantasmic character: 'The mythical recreation of the world from the deserted island gives way [in *Robinson Crusoe*] to the reconstitution of everyday bourgeois life from a reserve of capital. Everything is taken from the ship. Nothing is invented'.[3]Similarly, Deleuze points out, Jean Giraudoux's work, *Suzanne et le Pacifique*, offers a vision of the desert island as a transcendent, monadic territory, while unveiling its material impossibility. The desert island in this work functions as a double of Suzanne's Paris, 'a depository of ready-made, luxurious objects'. Giraudoux's island, for Deleuze, does not represent transcendent autonomy, but

1 A nascent version of the argument found in this chapter was presented at the 21st World Congress of the International Comparative Literature Association in Vienna, and published as Niall Sreenan, 'Dreaming of Islands: Individuality and Utopian Desire in Post-Darwinian Literature', *Island Studies Journal*, 12.2 (2017), pp. 267–80, doi:10.24043/isj.34.

2 Gilles Deleuze, 'Desert Islands', in *Desert Islands and Other Texts: 1953–1974*, ed. by David Lapoujade, trans. by Michael Taormina (Semiotext(e)/ MIT Press, 2004), pp. 9–14 (p. 10).

3 Ibid., p. 12.

derivation through repetition: a 'disarticulated' island derived from a larger body of land, 'a double without consistency, separated from the real'.[4] This doubling, therefore, shows the unreal island to have a pre-existing 'real' origin, to which it remains related by virtue of its repetition. Here, the desert island as a literary geographical space is a fantasy of transcendence, whose narrative representation seems to undermine its own ideational content.

If literary fantasies of desert islands stage a tension between dreams of transcendence and the unveiling of their illusory nature, physical, geographic islands underline the materialist dynamism that the desert island fantasy obscures. This can be emphasised, Deleuze implies, by a rudimentary understanding of geology and tectonics, which tell us how islands — which at the surface appear isolated, individuated, and autonomous — are processual and connected entities, always undergoing material processes of differentiation, emergence, and derivation from a larger but submerged earthly whole. A continental island becomes separated from a larger body of land, splitting off to take up a seemingly independent existence. An oceanic island rises from the ocean floor, surfacing from deep, a volcanic emergence that pierces the water's surface which obscures its volcanic origins. As such, both islands' conditions of possibility are what denies them transcendent finality. Islands, Deleuze writes:

> Reveal a profound opposition between ocean and land. Continental islands serve as a reminder that the sea is on top of the earth, taking advantage of the slightest sagging in the highest structure; oceanic islands, that the earth is still there, under the sea, gathering its strength to punch through to the surface.[5]

'These elements,' Deleuze continues, 'are in constant strife', their 'revulsion for one another' resulting in a constant differential exchange. The existence of an island is premised on the perpetual play between material forces, which ensures that it can never be said that an island exists, but only that it is relentlessly in the process of coming into and fading out of existence.

This analysis of geographical materiality in the figure of the desert island in the early part of Deleuze's intellectual life anticipates his later approach to reading Darwin's theory of evolution in *Difference and Repetition*. As I have already shown, in this later work Deleuze argues that Darwinian natural selection problematises a transcendent conception of taxonomic categories which impose a notionally timeless coherence on otherwise unstable biological existence by subordinating difference to conditions of resemblance and opposition. For Deleuze, Darwin's work does not deny the existence of taxonomic units but insists instead on their historical and biological contingency. Similarly, the island should not be taken as a sign of transcendence but a physical instantiation of the differential nature of geological and geographical materiality. What we perceive as notionally isolated units of biological life — individual organisms as well as species and variations — does not authorise a transcendent conception of biological existence, but 'on the contrary, [...] are understood on the basis of such fundamental mechanisms of

4 Ibid., p. 9.
5 Ibid.

natural selection as difference and the [physical actualisation] of difference'.[6] Like geological life, biological individuation is neither an illusion nor a transcendent entity; species and individuals are both real and constantly in motion, emerging and splitting off from a larger biological whole, to which they remain fundamentally connected. This conception of biological life can be found too in Elizabeth Grosz's work, who asks us to see life, species, and individual organisms, as 'a kind of contained dynamism [...] with a porous boundary, that feeds from and returns to the chaos which surrounds it'.[7] We can understand the island in the same way: a geo-oceanic embodiment of difference in itself, a 'contained dynamism', whose boundaries are contingent on their porousness, and whose existence depends on the possibility of its end.

In biographical and historical accounts of the genesis of Darwin's theory of natural selection, the meta-poetic figure of the island is an iconic and theoretically instrumental presence. As is well known, Darwin's visit to the Galápagos was an important waypoint in his intellectual formation and development of evolutionary naturalism. It is on these islands, twenty years prior to the publication of *On the Origin of Species*, that Darwin was struck by the singularity of the island species he observed as much as he was intrigued by their unmistakable resemblance to seemingly separate species on continental America. He speculates in his account of the *Beagle* voyage that this phenomenon of difference and resemblance was the result of a 'creative power', which acted 'according to the same law over a wide area'.[8] Later in *On the Origin of Species*, Darwin alludes to the significance of his time on the Galápagos, recalling that it was on these islands that he began to notice the inadequacy of a representational model of life, 'how entirely vague and arbitrary is the distinction between species and varieties', and to notice how the geographical distribution of similar species implied a 'deep organic bond, prevailing throughout space and time, over the same areas of land and water, and independent of their physical conditions' (*Origin*: 48). The problem of sameness and difference presented by these observations, Peter Bowler argues, was solved by Darwin via the realisation that islands are themselves a dynamic entity, porous zones of 'geographical isolation'.[9] Distinct species emerged on islands as a combination of biological evolutionary change (selection and reproduction) and the geological dynamism of island creation (emerging from the deep or disconnecting from a continent), where the island provides a separation necessary for the processes of biological change to produce insular forms of life that diverge more radically from those of their continental ancestors.

6 Deleuze, *Difference and Repetition*, p. 248.

7 Grosz, *Becoming Undone*, p. 27.

8 Charles Darwin, *Narrative of the Surveying Voyages of His Majesty's Ships Adventure and Beagle between the Years 1826 and 1836, Describing Their Examination of the Southern Shores of South America, and the Beagle's Circumnavigation of the Globe*, ed. by Robert Fitzroy, 3 vols (Henry Colburn, 1839), III (*Journal and Remarks, 1832–1836*), p. 474.

9 Peter J. Bowler, *Evolution: The History of an Idea*, rev. edn (University of California Press, 1989), pp. 162–63.

There is a suggestive dissonance, then, in Edward Larson's account of Darwin's development of the theory of natural selection, which attributes to Darwin's experience of the Galápagos Islands a force of creativity that approaches providential inspiration. Larson calls the islands 'Darwin's Eden', 'The Galápagos Wonderland', implying that they represent an epistemically liminal space between the irrational time of pre-Darwinian thought and a rational post-Darwinian world.[10] Julian Huxley writes that it was on the Galápagos that 'Darwin took the first step out of the fairyland of creationism into the coherent and comprehensible world of modern biology for it was here that he became fully convinced that species are not immutable'.[11] Devoid of materiality in these accounts, the Galápagos are assigned what amounts to a story of *ex nihilo* creation, a timeless mediating space between chaos and disorder, between the animal and the human. Darwin lands on the Galápagos like Crusoe to (re)create the world in his image. These narratives of the island origin of natural selection mirrors Ian Watt's influential study of the emergence of the English novel: the fantasy of island sovereignty and the creative force of the *homo economicus*, Watt argues, both nourish and are nourished by the historical emergence of a Puritan capitalist impulse in the late seventeenth and early eighteenth centuries in England. Crusoe can be read as the 'archetypal capitalist', whose vigorous individualism reflects that of the British empiricists of the seventeenth century, and whose reconstitution of a civilised world from a savage one dignifies Adam Smith's focus on the division of labour in *The Wealth of Nations* (1776).[12] This idea of the individual as a timeless, sovereign island is paradoxically reproduced by competitive social-Darwinian theories which dictate that natural selection further authorises, and naturalises, both individual, economic self-interest and competition. As Mike Hawkins suggests, Herbert Spencer's progressivist interpretation of Darwin's theory of biological evolution underwrote a pre-existing and popular strain of individualist economic liberalism in late nineteenth-century Britain which continues to be widespread today.[13] According to Hawkins, Spencer's evolutionism affirms that unlike biological organisms, the social organism is constituted of multiple centres of consciousness, the relations between which govern the progress of society as a whole. In other words, whereas Darwinian evolution suggests for Deleuze that the individual (whether figured as island or organism) is a durational, ephemeral entity, Spencer's paradigmatically capitalist interpretation uses evolution to reassign transcendent sovereignty to the individual, reproducing a paradoxical fantasy of an evolved but autonomous island.

In other accounts of Darwin's time on the Galápagos, scholars such as Bowler and Niles Eldredge affirm that the islands do symbolise a turning point in Darwin's

10 Edward J. Larson, *Evolution's Workshop: God and Science on the Galápagos Islands* (Basic Books, 2001), p. 3.

11 Julian Huxley, 'Charles Darwin: Galápagos and after', in *The Galápagos: Proceedings of the Symposia of the Galápagos International Scientific Project*, ed. by Robert I. Bowman (University of California Press, 1966), pp. 3–9 (p. 3).

12 Ian Watt, *The Rise of the Novel: Studies in Defoe, Richardson, and Fielding* (University of California Press, 1957), pp. 60–93.

13 Mike Hawkins, *Social Darwinism in European and American Thought, 1860–1945: Nature as Model and Nature as Threat* (Cambridge University Press, 1997), pp. 89–90.

thinking, not because the island is a transcendent space, but precisely because of its dynamic and porous nature.[14] The emergence and derivation of islands engenders what biologists call 'geographical speciation', facilitating the reproductive isolation of a group of organisms and allowing those organisms to diverge and to evolve into distinct but related species. In addition to the Galápagos Islands, Darwin also notices at this early stage of his career the manner in which the Irish and the English hare had diverged, owing to their island isolation and subsequent evolutionary bifurcation.[15] So for Darwin, in such accounts, the dynamism of islands — both the British and Galápagos archipelagos — and their differential relation to the dynamism of biological change, sharpens and makes visible those processes of selection, reproduction, and evolution, which over a less physically differentiated territory would be more difficult to perceive.

For Manuel DeLanda the process of evolutionary divergence that Darwin saw through the isolation of islands is an exemplary illustration of Deleuze's anti-essentialist conception of individuation.[16] The individuation of species in his account consists of two processes: the sorting operation of natural selection and the creation of reproductive islands which are effectively sealed off from the possibility of hybridity. This, he proposes, elucidates the fundamental reversal that Deleuze, through Darwin, applies to biological thought. The degree to which a species (which DeLanda understands as an 'individual') possesses an identity depends on the degree to which that species has become reproductively isolated from the differential network in which it is embedded. For some less-differentiated species, hybridisation remains possible and thus their identity is unclear; for others, they have in effect become islands, and their identity is almost but not entirely definitive. But as Darwin points out, the difference between a defined species and a hybrid variation is an arbitrary distinction; unless of course, as DeLanda suggests, that distinction is understood as inseparable from the historically contingent process of individuation that gives rise to it.

The figure of the island, then, symbolises and embodies contradictory tendencies in the afterlife of Darwin's theory of evolution and its application to wider social, economic, and philosophical fields. The island represents a figure through which to comprehend the differential, anti-essentialist naturalism that Deleuze, and DeLanda through Deleuze, reads in Darwin's theory of natural selection. It is also a historical, philosophical, and economic figure that enables a fantasy of individual liberty, autonomy, and agency into the service of which (as I have elaborated upon in Chapter 1) Darwin's work has been pressed. Deleuze's essay on desert islands suggests, however, that the latter interpretation of Darwinian evolution, the contradictory authorisation of transcendent biological autonomy, is fundamental not only to anti-materialist theories of economic individualism, but to a sense of the existential security of humanity. 'Humans cannot live, nor live in security,' Deleuze writes, 'unless they assume that the active struggle between the earth and

14 Bowler, *Evolution*, pp. 156–64; Niles Eldredge, *Darwin: Discovering the Tree of Life* (Norton, 2005), p. 39.

15 Darwin, *Narrative*, III, p. 245.

16 Manuel DeLanda, *Intensive Science and Virtual Philosophy* (Bloomsbury, 2013), pp. 42–43.

water is over, or at least contained'.[17] For Freud, in *Civilisation and Its Discontents*, the problem of civilisation is that it relies on a necessary but at the same time futile abjuration of humanity's inherent animality. Through Deleuze, we can say that the very concept of the human depends on a repeated literary, philosophical, and scientific reproduction of the fantasy of the desert island, the existence of which is based on the perpetual repression of the processual, dynamic nature of life which in Darwinian terms is coterminous with animality.

This chapter explores this tension in relation to Darwinian evolutionary theory by focusing on the literary representation of the island in three utopian novels centred on the theoretical and spatial figure of the island: Samuel Butler's *Erewhon* (1872), Aldous Huxley's *Island* (1962), and Michel Houellebecq's *La Possibilité d'une île* (2005). The novels span the nineteenth, twentieth, and twenty-first centuries, and each one can be read as an explicit response to Darwinian biology. Moreover, as this chapter will show, the spatial dynamics of these novels' imaginary islands can be related metaphorically and historically to the Darwinian conception of the human individual and its implications. If the fantasmic desert island is symptomatic of a submerged desire for human sovereignty, how does the utopian island function, where, at a generic level, the investment of a desire for human perfectibility (and thus a certain kind of finality) is constitutively explicit to the project of political and social dreaming? If the desert island is also an expression of desire to reproduce the conceptual split between humanity and animality, to what extent is the utopian literature of a post-Darwinian epoch invested in the same imaginative project?

The country of Erewhon in Butler's work is a distant, hermetic polity, encircled by a vertiginous mountain range in the interior of New Zealand. Its isolation has resulted in the evolution of a culture whose laws and customs appear opposed but uncannily similar to those of Butler's own nineteenth-century socio-political milieu. The novel's topographical rhetoric echoes the dynamics of island speciation in DeLanda's characterisation of Darwin's thought, where the radical divergence of a species is facilitated by reproductive isolation and, moreover, where an ancestral trace produces the resemblance of that species to nominally different, geographically distant groups of organisms. Butler's novel also undertakes an explicit examination of the ramifications of Darwin's theory of evolution in a section of the novel entitled 'The Book of the Machines'. Written by a scholar of Erewhon and quoted at length by the novel's narrator, this book explores the threat that machinery and industry would represent to humanity should the principle of biological evolution turn out to apply also to the progress of technology. Seeking to transcend the evolutionary threat of a rival species, in this case the non-human threat of technology, the book provides the theoretical basis on which the utopian inland island's autarchic isolation is instantiated.

Huxley's *Island* depicts a similarly hermetic island nation named Pala, whose remote oceanic location and autarchic social policy has facilitated the evolution of a spiritually elevated, peaceful, and economically efficient society. This work's engagement with Darwin's theory of evolution is made explicit in the novel's *récit*,

17 Deleuze, 'Desert Islands', p. 9.

which depicts a society whose prosperity is founded on the utilitarian application of evolutionary principles to all aspects of social, political, and economic life. Selective, eugenic breeding practices ensure the continued success of crops and agriculture as well as the biological improvement of the island's population, while a religious hybrid of Darwinian materialism and Buddhist stoicism engenders a collective, ecological conception of nature and distaste for individual self-interest.

The latter stages of Houellebecq's *The Possibility of an Island* depicts a distant, nominally Utopian future, in which the figure of the island is not merely a geographical one but both temporal and biological in nature too. The future that Houellebecq's work comes to describe is populated by so called 'neohumans', cloned, genetically modified human individuals, each of whom lives alone in hermetically sealed, isolated compounds, which protect them from the environmentally hostile external environment. Like Huxley's *Island*, the novel is, in part, an exploration of the Utopian possibilities of genomic technology and evolutionary biological theory: each neohuman lives hermetically and peacefully, dedicating his or her life to study and religious observance. But the novel also examines the negative ramifications of Darwinian thought, in particular the extension of the principle of the struggle to survive not only to the economic, but also the social, sexual, and affective aspects of human life. Most of the novel is narrated from the perspective of a twenty-first-century human, Daniel, whose acerbic and depressive autobiographical narrative describes the misery of living in a society in which individualist, neo-liberal capitalist competition and the principle of natural selection have come to dominate and degrade all forms of social relation. Seeking to mitigate and ultimately escape from the alienation of this social-Darwinian milieu, Daniel joins a techno-religious, Utopian cult whose research into genomics leads to the development of the hermetic, island mode of neohumanity, from whose future perspective the latter part of the novel is narrated.

Each of these three novels creates a topographical and textual site within which to examine the utopian socio-political and bio-technological possibilities of Darwin's theory of evolution. And in each of these work's narratives a sealed, hermetic island space can be understood to represent an impossible aspiration to social and biological autonomy. This chapter, therefore, postulates the significance of the spatial motif of the island, along with Zola's urban jungle and Hardy's rural wilderness, as a key site through which post-Darwinian literature responds to and retheorises the concept of the biological human. This chapter will go on to elaborate upon the representation of the island in these texts, the way they are bound up with a 'utopian impulse' towards the human individual and collective as an enclosed totality, through whose failure in each novel we can read a formally and thematically dialogic philosophical reflection on the similarly contradictory nature of the human-animal. At the same time, by establishing and exploring this tropic feature, this chapter also charts a new historical genealogy of the post-Darwinian novel, connecting Victorian anti-Darwinian satire, twentieth-century Utopianism, and contemporary French science fiction, via the common thread of their engagement with Darwin. It also connects notionally disparate generic traditions of the novel, realism and Utopianism, through the common vector of their response to Darwin. As such, this chapter can be read, like Chapters 1 and 2 on realist fiction, as combining a selective historical mapping of the literary reception of Darwin with a theoretically engaged analysis of the way these texts work to reconceptualise the human in the wake of the Darwinian shock to the anthropocentric *Weltanschauung*. In Zola's reception of Darwinian competition the human is understood to be caught between the nihilistic embrace of evolutionary violence and a delusional hope for revolutionary transcendence. In Hardy's works, this deadlock persists but represents the starting point for an affirmative but narratively uncertain engagement with the erotics of life. The three island-fictions with which this chapter is concerned ask whether the evolutionary autonomy denied to the human in Zola's and Hardy's works can be found in a Utopian projection of evolutionary futurity. In that respect, if in Zola's and Hardy's works the dream of a transcendent human existence is implicit, the works in this chapter explicitly elaborate a fantasy of human escape from evolutionary entanglement: the utopian island as purposeful zone of freedom from the vicissitudes of Darwinian animality.

The Conflicted Darwinisms of Butler, Huxley, and Houellebecq

In a recent study of Butler's life and works, David Gillott argues that to over-emphasise the influence of Darwinian evolution in Butler's literature under-estimates the significance of the evolutionary thought of Jean-Baptiste Lamarck.[18] For as well as a novelist and essayist, Butler was an art critic, a biblical critic, a painter, and wrote four works of evolutionary theory, *Life and Habit* (1878), *Evolution old and New* (1879), *Unconscious Memory* (1880), and *Luck, or Cunning* (1887), as well numerous essays on related subjects. For Gillot, Butler's *métier* is to argue against the professionalised, scientific orthodoxy of the Darwinian mode of evolutionary thought to propound an idiosyncratic, heterodox version of Lamarckian evolution, which stresses individual agency, skill, and the transmission of ancestral memory, as opposed to random variation and brutal selection. Butler, Gillott suggests, rejects Darwinian 'luck' in favour of Lamarckian 'cunning': life in Butler's view does not evolve as if mechanically driven onwards by random processes of variation and selection, but proceeds on the basis of the ingenuity and volition of the individual will.[19] Moreover, Butler sees the human organism as a cultural entity whose 'personal identity' exceeds the boundaries of the physical body, and sees non-physical entities like tools, texts, and ideas as 'extra-corporaneous limbs'.[20] Thus evolution for Butler is a cultural epigenetic process as well as a solely biological or somatic one. The organism and the species, he believes, evolves in dialogue with the environment which it inhabits and, following Lamarck, transmits 'acquired characteristics' developed in response to that environment to its progeny.

18 David Gillott, *Samuel Butler against the Professionals: Rethinking Lamarckism 1860–1900* (Legenda, 2015).
19 Ibid., p. 5.
20 Ibid., p. 9.

Gillott's analysis is aimed at correcting a critical 'commonplace' which tends to read Butler in relation to his fervent opposition to Darwinian natural selection, rather than through the positive project of a Lamarckian philosophy and aesthetics. The negative aspect of Butler's project is difficult to ignore since his critique of the nineteenth century's most prominent and influential evolutionary theorist, as well as being implicit in his adoption of Lamarckian concepts, takes place in a series of public ad hominem attacks on Darwin in the periodical press.[21] Yet it is argued that even the manner in which Butler repudiated Darwin was integrally Lamarckian. At the centre of Butler's dismissal of Darwinism lies a personal indictment of Darwin's failure as an author to acknowledge his antecedents in evolutionary theory. Butler's Lamarckism, then, seeks to correct this by championing Enlightenment theories of evolution, including those of Darwin's grandfather, Erasmus Darwin, as well as Lamarck. Butler's strategy, Gillott argues, is to emphasise the disingenuousness of Darwin's conceptual and literary methodology, since the latter's disavowal of his sources allows him also to disavow the importance of what he himself has inherited culturally, in Lamarckian fashion. This relational, constructivist ontology of author and text, nature and culture, allows Butler to attack Darwin's personal integrity, since his dishonesty is also scientifically insincere, and so by extension are his texts and theories.[22]

Gillott's argument that Butler's Lamarckism is largely unrecognised, upon which he builds his suggestive reading of Butler's conception of cultural and biological evolution, is not shared by all critics. Gillian Beer acknowledges Butler's Lamarckism, as do James Paradis and David Amigoni. Beer, however, frames Butler's singular attempt at an alternative form of evolutionary thought in terms of his self-confessed contrarianism which, she suggests, allows us to re-read Darwin's work with renewed energy.[23] As George Levine points out, latent in Butler's critical attitude, 'in an astonishingly anticipatory way', was a quasi-deconstructive ethos which stated that 'every proposition, nay every idea, carries within itself the seeds of its own undoing'.[24] Thus, to read Darwin through Butler, Levine suggests, is not to dismiss Darwin's theory of evolution entirely but to seize upon those contradictions and slippages in Darwin's thought and in that way to enrich it. This mode of reading, Amigoni notes, 'led Butler to the paradoxical conclusion that Darwin "was at no time a thorough-going Darwinian, but was throughout an unconscious Lamarckian, though ever anxious to conceal the fact from himself and his readers"'.[25]

21 James G. Paradis, 'The Butler-Darwin Biographical Controversy in the Victorian Periodical Press', in *Science Serialized: Representations of the Sciences in Nineteenth-Century Periodicals*, ed. by Geoffrey Cantor and Sally Shuttleworth (MIT Press, 2004), pp. 307–29.

22 Gillott, *Samuel Butler against the Professionals*, pp. 55–77.

23 Gillian Beer, 'Butler, Memory, and the Future', in *Samuel Butler, Victorian against the Grain: A Critical Overview*, ed. by James. G Paradis (University of Toronto Press, 2007), pp. 45–57 (p. 51).

24 George Levine, *Dying to Know: Scientific Epistemology and Narrative in Victorian England* (University of Chicago Press, 2002), p. 273; Samuel Butler, 'Life and Habit, Vol. 2', in *The Shrewsbury Edition of the Works of Samuel Butler*, ed. by Henry Festing Jones and A. T. Bartholomew, 20 vols (Dutton, 1923–26), XVIII, pp. 111–84 (p. 112).

25 David Amigoni, '"The Written Symbol Extends Infinitely": Samuel Butler and the Writing of Evolutionary Theory', in *Samuel Butler, Victorian against the Grain*, ed. by Paradis, pp. 91–112 (p.

Whether or not Darwin is an unconscious Lamarckian, or whether Butler's Lamarckism is ungraspable without an understanding of his anti-Darwinism, Amigoni affirms that what interests Butler fundamentally is an ontological paradox at the heart of the evolutionary concept of nature. In 'The Deadlock in Darwinism' (1890) Butler writes:

> Everything both is and is not. There is no such thing as strict identity between any two things in any two consecutive seconds. In strictness they are identical and yet not identical, so that in strictness they violate a fundamental rule of strictness — namely, that a thing shall never be itself and not itself at one and the same time.[26]

Butler's conception of the non-self-identical nature of biological materiality anticipates that of Deleuze, who through Darwin himself argues that organic life is inherently different — and differing — from itself. This identification of the immanently contradictory nature of evolutionary materiality and being, Amigoni argues, enables Butler to interrogate evolution through literary forms like satire and dialogue, and its implications in relation to identity, authorship, and culture. These ironising, dialectical forms allow Butler to address the inherently unstable, durational nature of literary modes to which self-contradiction is intrinsic.[27]

This way of understanding Butler's critical and literary methodology allows us to understand how even in his adoption of Lamarckian evolution, Butler is engaged with Darwin. Butler published an anonymous, philosophical dialogue entitled 'Darwin on the Origin of Species' in 1862, which was followed in 1863 by 'Darwin among the Machines', and in 1865 by '*Lucubratio Ebria*', the latter two of which provide important source material for *Erewhon*'s 'Book of the Machines'.[28] In his earliest essay about *On the Origin of Species*, Butler is supportive of Darwin's work, undertaking a complex analysis of the Malthusian aspect of natural selection and its implications for religious faith. In the latter two essays, he draws upon an analogy between natural selection in a biological context and technological evolution, wondering about the consequences for humanity should technology continue to evolve apace. In *Erewhon*, these speculative essays are given fictional treatment. The citizens of Erewhon have concluded that technological evolution is a threat to humanity and thus must be destroyed, and through this Butler plays upon the notion that evolution dissolves the 'strictness' of identity between objects, between human and machine, responding critically to the Darwinian affirmation of evolutionary descent. Just as his reading of Darwin leads him to envisage evolutionary futurity based upon 'what is' becoming something other than itself, Butler's relation to Darwin's thought is founded on an interpretative ethos which sees evolutionary thought carrying the seeds of its own transformation. Butler cannot be easily read

105); Samuel Butler, 'The Deadlock in Darwinism', in *Essays on Life, Art, and Science*, ed. by R. A. Streatfeild (A. C. Fifield, 1908), pp. 234–340 (p. 257).

26 Butler, 'The Deadlock in Darwinism', p. 243.

27 Amigoni, '"The Written Symbol Extends Infinitely"', p. 105.

28 Samuel Butler, 'Darwin among the Machines' and 'Lucubratio Ebria', both in *A First Year in Canterbury Settlement with Other Early Essays*, ed. by R. A. Streatfeild (A. C. Fifield, 1914), pp. 180–85, 186–95.

either as a solely Darwinian author or solely Lamarckian author, for his engagement with one is involved with the other. His radically eccentric evolutionism, by contrast, is instantiated in his singularly critical attitude and developed and explored, as this chapter will demonstrate, in his literary writing.

The question of whether Aldous Huxley is, unlike Butler, a straightforward disciple of the Darwinian model of evolution by natural selection and sexual selection is both simpler and more complex. Huxley's grandfather, T. H. Huxley, known also as 'Darwin's Bulldog', was one of Darwin's most fervent defenders and a committed evolutionist in his own right, whose legacy, R. S. Deese argues, neither Aldous nor his brother Julian, himself a prominent Darwinian evolutionist, could escape. The extent to which Darwinism, metonymically represented by T. H. Huxley, exerted a powerful influence over Aldous is emphasised by Deese who relates how, at the age of six, a young Aldous attended the unveiling by the Prince of Wales of a monumental statue dedicated to his eminent grandfather. Aldous, Deese writes, 'stood nervously to see the proceedings, and to gaze at this imposing statue through the throng of adults'.[29] For his own part, Huxley the novelist went on to describe himself as being 'in the tradition established by his grandfather', 'a cheerleader for evolution'.[30] So Aldous Huxley, unlike Butler, could in fact be read, as Thomas Hardy sometimes is, as a comparatively uncomplicated follower of Darwinian science, having been born under the shadow of its paternal law. However, Huxley did not simply champion evolutionary thought as a biological concept. He sought also, like Butler in his early essays, to examine its ethical and political implications in relation to human society. Most famously, his depiction of a eugenic dystopia in *Brave New World* (1932) can be read as a predictive indictment of the supposed utopian possibilities of Darwinian evolutionary thought in the hands of totalitarian political regimes. The portrayal in his late work *Ape and Essence* (1948) of a world destroyed by nuclear war is interspersed with vignettes of a society run by baboons condemn the notion of scientific and technological progress and the possibility of a regression to savagery that is latent in humanity's post-Darwinian condition.

For Peter Bowering, the primary influence of Darwinian evolution on Aldous Huxley came in the form of T. H. Huxley's assertion that the progress of evolutionary science would strip the world of spirit and spontaneity. Accordingly, Bowering writes, 'the debasement of man's moral nature by the increase of his knowledge' represents the central theme in Huxley's fiction.[31] Milton Birnbaum recognises the influence of Aldous's paternal grandfather and the body of thought of which he was the defender-in-chief, and extends the genealogical argument further. 'Huxley's soul,' Birnbaum writes, 'was always the battleground between

29 R. S. Deese, *We Are Amphibians: Julian and Aldous Huxley on the Future of Our Species* (University of California Press, 2014), p. 23.

30 Ibid., p. 5. Huxley is reported to have described himself as a 'cheerleader for evolution' in a conversation with Timothy Leary recorded in an unpublished document in possession of the U. S. Immigration and Naturalization Service: quoted in David King Dunaway, *Huxley in Hollywood* (Harper & Row, 1989), p. 354.

31 Peter Bowering, *Aldous Huxley: A Study of the Major Novels* (A. & C. Black, 2014), p. 25.

the challenging barks of "Darwin's Bulldog"' and the 'melancholy promptings' of his maternal great-uncle, Matthew Arnold. These two genealogical and paternal pressures meet in synthesis, Birnbaum says, in Aldous's exploration in his later years of the utopian possibilities of science and narcotics.[32] Ronald Sion echoes this reading, tracing the dominant thematic preoccupations of Huxley's fiction to Arnold's 'struggle for certainty in the late Victorian world of religious doubt', and to his grandfather's support for the scientific revolution that would cause this struggle.[33] Robert S. Baker identifies the culmination of these two competing exigencies in Huxley's *Brave New World*, whose dystopian vision of a eugenic future, he argues, is symptomatic of the post-Darwinian focus on 'process', but expresses an ambivalence about the consequences of that process for humankind.[34] This work depicts what, for Huxley, would be the oppressive, nightmarish consequence of humanity developing the capacity to control the process of evolution through eugenic engineering: a world where humanity is deprived of individuality, the capacity to think critically, and is coerced into submission by the state with hallucinogenic drugs and the opiate of popular entertainment.

If Butler's attitude towards Darwin and his drift towards an explicitly Lamarckian conception of cultural evolution is symptomatic of a general iconoclasm, Huxley's unorthodox view of the technological and societal ramifications of Darwin's theory could be said to originate in an Oedipal antagonism towards Darwinism. Like Butler, Huxley's rebellion against Darwinism seems to be defined by an anxiety of influence: his works' outward rejection of the techno-social possibilities of Darwinian thought is at the same time a way of exploring those possibilities. Butler's enactment of this desire took literary form: he examined Darwin's thought through fictional dialogue and speculative, creative essays, and criticised Darwin's own literary failings which he saw as symptomatic of the weakness of the evolutionary philosophy. Similarly, Huxley drew upon the possibilities of literary form: the dystopian fiction of *Brave New World* and the Utopian fiction of *Island* are spaces in which to interrogate both the destructive and the redemptive possibilities of Darwin's philosophy.

As Beer suggests, it is difficult to find a contemporary writer who is not either — consciously or unconsciously — Darwinian, since Darwinism, in one form or another, has passed into the cultural shadows of common sense. However, Michel Houellebecq is certainly more conscious than others of the legacy of Darwinian thought, at least as far its literary and cultural legacy is concerned. In his early novel *Les Particules élémentaires* [*Atomised*] (1998), one of the novel's protagonists, the sexually obsessive misanthrope, Bruno, encounters the work of Julian Huxley, given to him by his half-brother, a molecular biologist:

32 Milton Birnbaum, *Aldous Huxley: A Quest for Values* (Transaction, 2005), p. 172.

33 Ronald T. Sion, *Aldous Huxley and the Search for Meaning: A Study of the Eleven Novels* (McFarland, 2010), p. 14.

34 Robert S. Baker, 'Brave New World: Huxley's Dystopian Dilemma', in *Aldous Huxley's 'Brave New World'* (Chelsea House, 2003), pp. 93–104 (p. 93).

> Michel se leva, sortit de sa bibliothèque un volume intitulé *Ce que j'ose penser.* 'Il a été écrit par Julian Huxley, le frère aîné d'Aldous, et publié dès 1931, un an avant *Le Meilleur des mondes.* On y trouve suggérées toutes les idées sur le contrôle génétique et l'amélioration des espèces, y compris de l'espèce humaine, qui sont mises en pratique par son frère dans le roman. Tout cela y est présenté, sans ambiguïté, comme un but souhaitable, vers lequel il faut tendre.'[35]
>
> [Michel went over to the bookshelf and took down *What Dare I Think?* and handed it to Bruno. 'It was written by Julian Huxley, Aldous's older brother, and published in 1931, a year before *Brave New World.* All of the ideas his brother used in the novel: genetic manipulation and 'improving the species' — including the human species — are here. All of the ideas are presented as something which society should strive for.']

For Bruno, the vision offered by *Brave New World* only came to be seen as dystopian after Nazism compromised eugenic thought and the Huxleys (Julian and Aldous) sought to distance themselves from its negative connotations. Whether we should take Bruno's reading seriously or not is questionable: yet, in putting its finger on an ambivalence towards evolutionary thought in Huxley's work, Houellebecq alerts us to a similar complexity in his own. For Victoria Best and Martin Crowley, Houellebecq is certainly a materialist, but one 'who is infinitely depressed by materialism', both committed to a broadly Darwinian, naturalist conception of nature and society and simultaneously appalled at the suffering, the violence, and the despair it anticipates for human life.[36] Houellebecq's fiction, John McCann argues, takes place in a Darwinian universe, where a Schopenhauerian conception of the struggle to survive is extended to the emotional, economic, and social lives of humans.[37] In Houellebecq's fictional universe, McCann says, the anti-teleological nature of evolution is a malign force: '[e]volution is neither goal-directed nor a process of growth that realizes some potential essence. It proceeds by natural selection, the means by which the fittest survive. The struggle eliminates those who do not fit into the circumstances'.[38] I have, as have other critics, underlined the way in which, as in Zola's work, the dynamic of evolutionary struggle in Houellebecq's work connects capitalist competition and biological life, focusing in particular on neoliberal biopolitics.[39] Ben Jeffery couches this preoccupation with struggle in slightly different terms, locating Houellebecq's depressive materialism in his reading of H. P. Lovecraft's malignant worldview.[40] 'Humans are just animals', Jeffery

35 Michel Houellebecq, *Les Particules élémentaires* (Flammarion, 1998), p. 192; *Atomised*, trans. by Frank Wynne (Vintage, 2001), p. 189.

36 Victoria Best and Martin Crowley, *The New Pornographies: Explicit Sex in Recent French Fiction and Film* (Manchester University Press, 2007), p. 209.

37 John McCann, *Michel Houellebecq: Author of Our Times* (Peter Lang, 2010).

38 Ibid., p. 8.

39 Niall Sreenan, 'Universal, Acid: Houellebecq's Clones and the Evolution of Humanity', *Modern & Contemporary France*, 27.1 (2019), pp. 77–93, doi:10.1080/09639489.2018.1557130; Carole Sweeney, *Michel Houellebecq and the Literature of Despair* (A. & C. Black, 2013), p. ix.

40 Houellebecq reads Lovecraft's depiction of abject, monstrous, alien beings as symbolic for the disgust he felt for the world as well as symptomatic of a need to escape it: Michel Houellebecq, *H. P. Lovecraft: Against the World, against Life*, trans. by Dorna Khazeni (Gollancz, 2008). Ben Jeffery, *Anti-Matter: Michel Houellebecq and Depressive Realism* (Zero Books, 2011), pp. 6–8.

writes of Houellebecq's fictional universe, and that makes 'self-interest paramount'. Even as Houellebecq's characters are incapable of anything other than absolute, instinctual narcissism, they are also aware 'with some horror' of 'the essential unsustainability of individualism'.[41] Others have noted the connection between Houellebecq's work with that of Zola. However, as Louis Betty argues, where Zola's method is to 'confront his characters with the inexorabilities of their biological and environmental conditioning', Houellebecq's method is thematically naturalistic but in addition has a specifically religious or metaphysical purpose.[42] 'Heredity and environment are not [...] Houellebecq's central concerns; he is interested in God's absence and the submission to matter that such absence demands, which deprives human life of a meaning that might escape its immediate conditioning.'[43] Here again is an expression of Houellebecq's contradictory materialism: hopelessly stuck between a faith in materialism, naturalism, and realism, and its resultant despair.

It is clear that Darwin's theory of evolution and natural selection in particular, and its broad philosophical, sociological, technological, and economic implications, have exercised significant influence over each of these three authors. Butler, Huxley, and Houellebecq, along with Zola and Hardy whose works I have analysed in my previous chapters, belong to an important genealogy of prominent British and French novelists for whom naturalism, science, and Darwinian thought are thematically crucial to their fiction. And as with the complexity of Zola's and Hardy's reception of Darwin, which anticipates their works' response to the Darwinian revolution, the conflicting nature of the reception of Darwin by Butler, Huxley, and Houellebecq prefigures a contradictory dynamic in each of their visions of post-Darwinian utopian humanity. Butler's career can be crudely divided into a pro- and anti-Darwinian period, the latter being inseparable from the former; Huxley is divided by competing genealogical and philosophical exigencies, being both wedded to Darwinian naturalism and sceptical of its Utopian implications; and Houellebecq is a paradoxical Darwinian, critical of the consequences of the encroachment of naturalism and materialism on human relations, but seemingly pessimistic about the possibility of resisting that process. The same sense of division and paradox in relation to Darwinian evolution can be found in their novels, which elaborate fantasies about its utopian possibilities which are, in turn, haunted by the dangers those possibilities bring about. More fundamentally, however, something about Darwin's work invites in each of these authors an imaginative desire. Unlike Zola and Hardy, this desire is not the realist urge for totalising representation, but a utopian aspiration. We might call it, after Fredric Jameson and Ernst Bloch, a 'utopian impulse', where thinking is a form of 'venturing beyond' towards something 'Not-Yet-Conscious, Not-Yet-Become'.[44] The following section traces

41 Jeffery, *Anti-Matter*, p. 15.

42 Louis Betty, *Without God: Michel Houellebecq and Materialist Horror* (Penn State University Press, 2016), p. 21. Sandrine Rabboseau, 'Houellebecq ou le renouveau du roman experimental', in *Michel Houellebecq sous la loupe*, ed. by Murielle Lucie Clément and Sabine van Wesemael (Rodopi, 2007), pp. 43–51.

43 Betty, *Without God*, p. 21.

44 Ernst Bloch, *The Principle of Hope*, trans. by Neville Plaice, Stephen Plaice and Paul Knight, 3 vols (MIT Press, 1995), I, p. 6; Jameson, *Archaeologies of the Future*.

this desire across these three novels, exploring the way it draws on Darwin shattering of sovereign humanity, to reconstitute the possibility a post-Darwinian future for humanity nevertheless, focusing in particular on the spatial valences of utopian islands.

The Utopian Impulse

Butler's novel signals its intertextual engagement with the Utopian literary tradition in its title, which is also the name of the utopian country it depicts: 'Erewhon' is an inversion and a distortion of the English translation of the neologistic title of Thomas More's originary work *Utopia* (1516), signifying 'nowhere', *Ou-topia* being an Ancient Greek portmanteau meaning 'no-place'. More's *Utopia* takes the form of a fictional travelogue, narrating in its second book the discovery by the philosopher-sailor, Rafael Hythloday, of a peaceful, socio-politically ideal island-nation at the margins of a colonial frontier.[45] Similarly, Butler's text is narrated by an unnamed traveller who retrospectively recounts his journey to the margins of British geographical imagination, this time in a nineteenth-century imperial context. Arriving in New Zealand and travelling to its uncharted interior in search of new land, the novel's protagonist fortuitously discovers a geographically, politically, and culturally autarchic nation, the splendour of which initially promises the aesthetic perfection of humanity. Echoing More's description of the architectural beauty of Utopia and its capital city in particular, with its combination of broad streets, stately arches, and 'well-ordered and fruitful gardens', Butler's narrator is struck by the beauty of Erewhon, the capital city of which combines 'majestic towers', arcades, and marble statues with pastoral gardens replete with an abundance of birds and fruit, giving the narrator 'the impression of great peace and plenty' (*Erewhon*: 138).[46] *Erewhon*'s title does not merely allude to its engagement with More's *Utopia* but mirrors some of the sixteenth-century text's thematic concerns regarding order and perfection.

Huxley's novel *Island* is a formal and narrative echo of both these texts. The novel is narrated by Will Farnaby, a British journalist, who relates his travels to the remote and previously impenetrable, hermetic island nation of Pala. Just as *Erewhon*'s narrator travels in search of colonial wealth, Farnaby's aim in travelling to Pala is to facilitate the usurpation of its leader by a neighbouring nation, the exploitation of its natural oil reserves, and the effective colonisation of the island. And just as *Erewhon*'s narrator and More's Hythloday discover places of utopian and aesthetic perfection, Farnaby finds on the island of Pala a population of 'exquisite' men, women, and children, who inhabit a country of Edenic, pastoral plenitude. As Farnaby is given a tour of the island he marvels at the splendour of its built environment and the ingenuity of its agricultural landscape: 'Nature here was no longer merely natural', Farnaby says of a series of terraces for rice paddies, houses, and boundary walls that

45 Thomas More, *Utopia*, ed. by George M. Logan and Robert M. Adams, trans. by Robert M. Adams (Cambridge University Press, 1989).
46 Ibid., pp. 45–46.

ascend the slope of a mountain range, 'the landscape had been composed, had been reduced to its geometrical essences, and rendered, by what in a painter would have been a miracle of virtuosity' (*Island*: 25). Where in Zola's and Hardy's works, the aesthetic — the novel, the painting, or the village dance — represent possible forms of momentary or speculative escape from the cruelty of biological materialism, here, the utopian civilisation is itself a kind of aesthetic project. Indeed, while in both More's *Utopia* and Butler's *Erewhon* nature and civilisation have been wrought in frictionless synthesis to create utopian perfection, on Pala nature and culture no longer seem opposed but fixed in a dynamic of complementary mutuality.

The homology between Huxley's *Island* and Butler's *Erewhon* is acknowledged by Huxley, with an intertextual reference to Butler's work. Although the island of Pala is understood to be inaccessibly remote and surrounded by impassable and treacherous seas, Farnaby does manage to access the island. As his boat flounders off the coast of Pala, Farnaby first finds himself shipwrecked on the island, only to survive a fall from a cliff as well as an encounter with a deadly snake. Commenting on the miraculous nature of his arrival on the island, Farnaby recalls Butler's novel, and offers a quotation from its protagonist, who himself marvels at his luck in discovering a beautiful and apparently perfect society: '"Remember the beginning of Erewhon" he said. "As luck would have it, Providence was on my side"' (*Island*: 22). As well as signalling the genealogical continuity between these two texts, this meta-fictional gesture indexes the spatially ironic dimension in utopian literary discourse. Farnaby's comments here not only on his luck at finding such an exquisitely prepossessing people and country, but also on the fortuitous nature of penetrating a physically and theoretically enclosed, autarchic space. The narrator of *Erewhon* makes an identical observation, from which *Island* quotes, when breaching the seemingly impenetrable borders of the country of Erewhon. Although not an island in the oceanic sense, Erewhon is encircled by a mountain range which seals the country off from its external environment, making it in effect an inland *insula*. Erewhon's inhabitants actively police its borders with a set of grotesque, threatening statues, designed to ward off potential interlopers, just as the Palanese rigorously police their borders, supplementing their natural oceanic boundaries with the summary deportation of those they deem 'undesirable' to the island's broader cultural ethos. Genealogical intertextual connectivity links these islands and prefigures the topographical contradiction inherent in each of these narratives: in the same moment that the novels' narrators penetrate and describe the islands of Pala and Erewhon, their autonomy from the world is laid bare to narrative representation and through that infiltration is undermined.

In *Utopics*, Louis Marin explores the relation between Utopian narratives and the semiotic valences of their topographical features. He argues that More's island of Utopia is not a 'natural' discovery but is constituted by a deliberate act of enclosure, a 'pure act of creation'.[47] Utopia is created, Marin notes, by its king and creator, Utopus, by removing the isthmus that once connected the country

47 Louis Marin, *Utopics: The Semiological Play of Textual Spaces*, trans. by Robert Vollrath (Humanities Press International, 1990), pp. 105–06.

previously known as 'Abraxas' to the mainland. In other words, Utopia is not a pre-existing place, a deserted island waiting to be discovered, but an example of what Deleuze would call a 'continental island'. Yet, unlike the natural splitting-off that constitutes the geographical island, Utopia's existence is brought about by an act of deliberate self-isolation through the destruction of a continental connection and the constitution of new island boundaries. For Marin, the passage from Abraxas to Utopia, from the chaos of undifferentiated, continuous space to the order of differentiated, discontinuous plural spaces, represents nothing less than the passage from nature to culture, from the animal 'to the true state of humanity'.[48] Utopian enclosure is a form of incision through which the human makes itself separate from that which constitutes it, instituting a boundary between humanity and the animal, between civilisation and savagery. Similarly, for Deleuze in 'Desert Islands' humanity's disavowal of the differential dynamics through desert island creation is also a Utopian act of inscription, marking out nothing other than the fantasy of transcendent autonomy on which the idea of the human is based, and to which it appeals.

Fredric Jameson's theorisation of Utopian discourse accords with the argument that geographic, political Utopian enclosure and humanity's act of ontological self-creation are complementary. He implicitly elaborates upon Deleuze's notion that desert islands, and utopian spaces, originate in a foundational human fantasy of autonomy. 'The Utopian', Jameson argues, is practically an unlimited category; to the tradition of explicitly utopian literature and social programmes or communities, we can also add fantasy, advertising, revolutionary political theory, consumer products, and any other cultural form. Yet, common to both the explicitly and implicitly utopian project is a 'commitment to closure (and thereby to totality)'.[49] Thus utopian desire is expressed unconsciously in unexpected and banal non-literary or ostensibly a-political objects and practices. He sees in the most prosaic elements of contemporary culture a utopian unconscious which is defined by a commitment to enacting closure and the construction of definitive boundaries of individual difference. 'Even the most subordinate and shamefaced products of everyday life, such as aspirins, laxatives and deodorants, organ transplants and plastic surgery' all offer the prospect of an enclosed, finalised body.[50] These quotidian, corporeal forms of Utopian enclosure reaffirm that what the creation of a desert island, 'Utopic' space, seeks to achieve is the constitution of a pure and definitively autonomous human, in both individual and collective terms.

The 'islands' of Erewhon and Pala are exemplary Utopian spaces in Jameson's and Marin's terms. Both countries are couched in naturally isolated, remote locations, the one bounded by a seemingly impassable mountain range, the other by treacherous seas. Yet, to maintain these boundaries, both countries deliberately seek to maintain and finalise that natural enclosure, enacting what Marin calls 'Utopic inscription', seeking, as Jameson suggests, definitively to close themselves

48 Ibid., p. 106.
49 Jameson, *Archaeologies of the Future*, p. 4.
50 Ibid., p. 6.

off and thereby to recreate themselves as a form of finalised totality. As such, the incursions by Farnaby and *Erewhon*'s narrator only reconfirm the constructed spatial coordinates of the utopian island. For these islands to sustain a Utopic disconnection from their surroundings, their acts of repudiation must be carried out repeatedly, or else interlopers such Farnaby and his literary predecessor from *Erewhon* undermine their isolation. The complementarity of nature and culture that exists on these islands, therefore, is achieved not through a simple synthesis of the two but, paradoxically, through violent human acts of organised exclusion: Erewhon's and Pala's perfection and plenitude is premised on negation, on their capacity to delimit access to their walled cities, and the corresponding organisation of nature's chaos into walled gardens and agricultural landscapes. The harmonious co-existence of nature and culture is premised on an act of Utopic inscription, the intention of which is to excise the human from nature and thus enable the human to dominate it. In Deleuze's terms, these two islands, by seeking to abjure all relations with the external world, by attempting to render themselves discontinuous from the continuous space of the land or the ocean, make an attempt at material, political, and geographical enactments of the *dream* of an island: expressions of humanity's fundamental narcissism, the fantasy of being already separate from and superior to the natural world from which it derived.

The specifically utopian drive to human autonomy and its expression in geo-political, corporeal, and biological guises is explored in Houellebecq's *La Possibilité d'une île*. Echoing the central narrative trope of classical utopian fiction, this novel delineates in part the journey of its protagonist, Daniel, to an island utopia — the journey from savagery to a form of notionally superior humanity. The twenty-first-century iteration of Daniel, who in the novel is designated simply by the name 'Daniel1' is a cynical, nihilistic, and provocative French comedian and screen-writer. His autobiographical narrative casts Western society as an affective desert, a savagely competitive and uncaring milieu where the nihilistic embrace of narcotic self-destruction, sexual hedonism, consumerism, and an attitude of ironic distance together represent the only reparative possibilities in an otherwise barbarous world. In contrast, the Elohimite cult that Daniel1 joins promises redemption through immortality. Yet this offer is only partly understood as a spiritual achievement and is presented largely as a material, technological possibility enabled by research into genomic technology, human cloning, and genetic modification. Daniel1's journey then, unlike that of Farnaby, *Erewhon*'s unnamed narrator, and *Utopia's* Rafael Hythloday, is not geographical but temporal. He commits himself to the Elohimite programme of genomic immortality and the second half of the novel, narrated from the perspective of two of Daniel1's cloned descendants, Daniel24 and Daniel25, details the Utopian conditions of isolation in which Daniel1's new iterations live.

Together, the neohuman individual and the compound in which that neohuman lives represent a dual form of Utopic enclosure. The narratives of Daniel24 and Daniel25 describe a world riven by cataclysmic environmental collapse and overrun by small groups of non-cloned humans, whose lives, in opposition to that of the neohumans, seem barbarous and anachronistic. The neohuman race, including

Daniel1's neohuman iterations, live in hermetically sealed, environmentally controlled compounds which protect their inhabitants from the external world. Each compound is fitted with its own electric generator which powers its protective, electric fence to exclude the non-neohumans and, moreover, the external walls of Daniel24 and 25's 'residence' are coated with a layer of 'une petite quantité de radium, à la radioactivité atténuée, qui protégeait efficacement' [slightly radioactive radium, which gave effective protection] from their climatically hostile external environment (*Possibilité*: 181; *Possibility*: 152). The neohuman too is itself a form of utopian island. They do not require food or water to nourish themselves; they have been genetically modified to derive nutrition autotrophically, from the sun as well as from the mineral salts generated by a machine within each compound. The neohuman has dispensed with the need to introject and excrete any physical matter, constituting itself as a physically closed system. And, as if to protect against the kind of erotic reproductive relation we find in Hardy's response to Darwinian sexual selection, neohuman reproduction is not dependent on sexual dimorphism; at the expiry of each clone, a newly produced body is delivered to the compound from a centralised cloning centre, to take up the identity of the last clone and that of each of its predecessors, theoretically projecting the personhood of Daniel1 into an abstract timelessness. So the body of the neohuman is a biological projection of the compound itself: a sealed, sovereign, and timeless monad, excised from the world in which it lives, to which it no longer has any relation except through repudiation. The human has become, in effect, a material instantiation of the fantasmic desert island. In this state of pure biological and social autarchy, the neohumans live in seemingly complete and untroubled serenity, in which neither the joy nor the misery of being mortal and relating to others penetrates a state of hermetic isolation.

Formally speaking, Houellebecq's novel echoes the dialogic narrative structure of More's *Utopia*. The first section of More's work, the 'Dialogue of Counsel', is made up of a series of discussions on the ills of contemporary, fifteenth-century Europe between More in a fictional guise and the sailor-philosopher Rafael Hythloday, as well as on various aspects of political philosophy. The second book, 'Discourse on Utopia', consists in the detailed elaboration of the ideal state of the Utopian commonwealth, as related to More by Hythloday, and subsequently narrated by More to the reader. Similarly, Houellebecq's work is divided into two distinct parts. The first book, Daniel1's narrative, diagnoses the ills of the cultural, political wasteland of Europe as he perceives it, and is interpolated with numerous discussions between Daniel1 and other characters on political philosophy, religion, and evolutionary theory, a number of which have been rehearsed within Daniel1's various comedy performances, films, and other writings. The second book, narrated by Daniel24 and, on his expiry, Daniel25, describes post-human life and the conditions which sustain and make it possible. As George Logan and Robert Adams note in their introduction to More's work, the relation between the two books of *Utopia* is a dialogic one, the 'analysis of the evils of the existing society' forming an 'appropriate prelude to a discussion of a possibly better one'.[51] It is in

51 George M. Logan and Robert M. Adams, 'Introduction', in More, *Utopia*, pp. xi–xxix (p. xxi).

this juxtaposition of these two books and, by extension, the juxtaposition of Europe and Utopia, that sharpens what is distinctive about each. Houellebecq's work echoes and complicates this dialogic structure, by dispensing entirely with a linear passage from the first book to the second, from Europe to Utopia, by formally intertwining their two narratives.

This formal conceit in Houellebecq's work is integral to the novel's narrative and can be read as a structural enactment of the tension between humanity's desire for autonomy and its individual and collective reliance on a relational social, biological, and political ecology. The Elohimites, for all their technological advancement, do not succeed in developing the capacity to download human consciousness and in that way transmit it seamlessly from one clone to the next. Instead, 'Book 1', Daniel1's autobiographical account of his life, his distaste for the world, his increasing depression, and his drift towards the Elohimite cult, represents for the neohumans an image of his consciousness. 'Book 2' describes the lives of Daniel1's clones who, in lieu of being able to digitally transmit consciousness across generations, are dedicated instead to reading Daniel1's autobiographical testimony, in response to which they compose their own exegetical commentary, thus creating a series of related intertextual commentaries. The novel continuously oscillates between the individual narratives and testimonies of Daniel1, 24, and 25, each of which is implicated in the other, setting up a dialogic configuration between times, consciousnesses, and texts. Upon its arrival, a newly delivered neohuman takes up the endeavour of reading and commenting upon the life-stories and commentaries of each of his predecessors, with the intention of extending Daniel1's consciousness indefinitely. Houellebecq's novel formalises the impossibility of that attempt, however, by grounding the identity of each neohuman clone in an individual text, each of which is constituted by its relation to the texts in response to which it is composed. The theoretical urtext of Daniel1's consciousness, therefore, is revealed to be composed of a series of internal dialogues, whose extension into futurity is constituted by an imperfect form of interpretive transmission. Daniel1's conscious is not a desert island, it is not a homogenous, unchanging entity extended into the future. Rather, it is a dynamic network of multiple interrelated consciousnesses, whose dialogue with one another engenders a continual transformation of that entity.

The genealogical, intertextual relations between these three novels is itself an instance of this tension. Each novel is both an autonomous text as well as a text written in response to More's *Utopia* specifically, and to the Utopian literary tradition generally. Moreover, each novel can be read in relation to the other; as intertexts in the case of *Erewhon* and *Island*, but also as exegetical commentaries on one another, the relations between each text constituting a literary network or genealogy of Utopian literature. Houellebecq's neohuman/human dialogue is a commentary on how the tradition of Utopian literature — and by implication literary art in general — proceeds, paradoxically, on the basis of originality grounded in repetition. This point also raises the issue of genre and the extent to which, in opposition to the Jamesonian conception of 'the Utopian' as a broader, multi-medial praxis that includes political thought and everyday life, these types of novels constitute a distinctive kind of Utopian discourse. Perhaps it is less productive to read these novels as part of a distinctive Utopian genre than in terms of their satirical function. As Dominic Baker-Smith points out, contemporary discussions about *Utopia* oscillate uneasily between taking seriously the apparent political and philosophical programme of More's work and taking seriously his humour and wit by regarding the work as a 'sophisticated game, a *jeu d'esprit*'.[52] More's *Utopia* is an affirmative negation of the political situation of sixteenth-century England and an uncanny mirror of it, possessing many of the same geographical and political features. This ironic uncertainty, however, can be regarded as an integral part of this work's effect, eliciting in readers a hermeneutic uncertainty, a readerly attentiveness which alerts the reader to their own interpretive decision-making.

Similarly, readers of Butler's *Erewhon* have frequently emphasised the satirical intentions of the novel as well as the ironic, satirical tenor of Butler's *œuvre* as a whole. Lewis Mumford's introduction to the 1927 Modern Library edition of *Erewhon* argues for the satirical nature of the text and rejects its association with Utopianism.[53] James Paradis classifies *Erewhon*, along with his later work *The Way of All Flesh* (1903), a bitingly humorous autobiographical novel that targets the stultifying social conventions of Victorian society, as 'two culture-probing satires'.[54] Other critics have affirmed that it is through the lens of satire and not Utopianism or the Utopian literary tradition that Butler's work has usually been understood by critics.[55] Yet, as is implied by the novel's title, and in the 'recurrent trope of reversal' throughout the novel, the key to reading this work might not be to see it as absolutely Utopian nor entirely satirical, but as a satire of Utopian literature and by extension the impulse that it appears to depict.[56] What Butler's protagonist finds in the country of Erewhon is not a Utopian paradise, as he initially believes, but a society that is ruled by a venal elite, where illness is punishable as a crime, where financial corruption is treated with empathy, and where religion and faith have been superseded by the worship of money. He discovers that Erewhon's universities, 'The Colleges of Unreason', are decadent, anachronistic institutions dedicated solely to the study of 'hypothetics'; that Erewhon's financial institutions, the 'Musical Banks', are empty, purely symbolic entities where the Erewhonians visit to indulge in religious self-deception. In other words, what the narrator finds in Erewhon is not a Utopian negation of the Victorian society from which he came, but an inversion

52 Dominic Baker-Smith, *More's Utopia* (University of Toronto Press, 2000), p. 39.

53 Lewis Mumford, 'Introduction', in Samuel Butler, *Erewhon, and Erewhon Revisited* (New York: Modern Library, 1927), pp. xix–xxvii.

54 James G. Paradis, 'Introduction', in *Samuel Butler, Victorian against the Grain*, ed. by Paradis, pp. 1–18 (p. 1).

55 Joshua A. Gooch, 'Figures of Nineteenth-Century Biopower in Samuel Butler's *Erewhon*', *Nineteenth-Century Contexts*, 36.1 (2014), 53–71 (pp. 54–57), doi:10.1080/08905495.2013.867599; Sue Zemka, '"Erewhon" and the End of Utopian Humanism', *ELH*, 69.2 (2002), pp. 439–72 (p. 439).

56 Ann-Barbara Graff argues that Utopian satire is a distinctively Victorian trope that arises out of the ethical concerns with futurity suggested by evolutionary science: Ann-Barbara Graff, '"Administrative Nihilism": Evolution, Ethics and Victorian Utopian Satire', *Utopian Studies*, 12.2 (2001), pp. 33–52.

of the expectations of the reader of Utopian fiction, such that the Utopian island represents nothing other than a mirror of Victorian society's worst elements. This 'devious trick', writes Sue Zemka, means that '*Gulliver's Travels* and not *Utopia* is the true template for [*Erewhon*]'.[57] *Erewhon* can be read as a satire of Utopian literature's conventions and as a commentary on Victorian society by revealing the strange delusions of a foreign island to be those of the very public that reads the novel.[58]

Critical responses to Huxley's *Island* note its own satirical intentions and the manner in which, like Butler's work, the novel wreathes itself in ironic reversals. Frank Kermode calls *Island* 'one of the worst novels ever written' and at least one critical review seems to have dismissed it on account of its apparently impoverished Utopian dogmatism.[59]However, others have argued that the novel's didacticism, philosophical incoherence, and uneven plotting should be read as functions of its status as a text in the 'non-Leavisonian' critical tradition, according to which its function is to instigate thought through satirical negation rather than to effect, narrate, or describe.[60] For Robert Elliot, *Island*'s Utopianism is a repudiation of a hegemonic ideological cynicism. Although the islanders themselves purport to reject the ideological positions of both of capitalism and communism, and can themselves be read as resisting ideological Utopianism, Elliot argues that their post-ideological optimism is itself a negation of anti-Utopian pessimism. For that reason, *Island* can be read as an attempt 'to [speak] cogently against despair', the despair of anti-Utopianism, while at the same time warning against the consequences of that optimism turning into blind, ideological dogmatism.[61]

Huxley's work does not engage in the kind of narrative, generic, or satirical inversions that Butler's work does. However, its apparent earnestness in seeking to portray a Utopian world, in contrast to the strikingly dystopian picture he constructs in *A Brave New World*, to which *Island* can be read as a response, poses the question of how to read the latter. Simon Dentith, commenting on *Erewhon*, argues that the hermeneutic difficulties that the combination of Utopian and satiric writing poses, the shifting of positions through multiple layers of negation, prompts 'alert reading'.[62] Huxley's island urges a similar alertness, inviting the reader to question the sincerity of its didacticism, while at the same resisting any over-determined interpretation. Similarly, the problem that irony, satire, and Utopian negation poses to hermeneutic certainty is perhaps the most distinctive aspect of

57 Zemka, '"Erewhon" and the End of Utopian Humanism', p. 439.

58 It is of course possible to read this satirical move in More's inaugurating text too. However, there is significant scholarly disagreement as to whether this kind of reading is justified. See A. R. Heiserman, 'Satire in the Utopia', *PMLA*, 78.3 (1963), pp. 163–74, doi:10.2307/460858.

59 Frank Kermode, '"Island" by Aldous Huxley', *The Partisan Review*, 29.3 (1962), pp. 472–73 (p. 472); William Barrett, 'Atlantic Bookshelf', *Atlantic Monthly*, April 1962, pp. 150–60.

60 Wayne Booth includes *Candide, Rasselas*, and *Gulliver's Travels* in this grouping: Wayne Booth, 'Yes, But Are They Really Novels?', *Yale Review*, 51.4 (1962), pp. 630–36 (p. 631).

61 Robert C. Elliott, *The Shape of Utopia: Studies in a Literary Genre* [1970], ed. by Phillip E. Wegner (Peter Lang, 2013), p. 97.

62 Simon Dentith, 'Imagination and Inversion in Nineteenth-Century Utopian Writing', in *Anticipations: Essays on Early Science Fiction and Its Precursors*, ed. by David Seed (Syracuse University Press, 1995), pp. 137–52 (p. 139).

Houellebecq's entire *œuvre* as a novelist. As Martin Crowley and Victoria Best argue, Houellebecq's work seems dedicated to 'essaying a complete picture of the contemporary world (though more in the mode of satire than of realism)'. They see Houellebecq as ethically concerned with offering a critique of the hegemony of the so-called free market, liberal economics, as well as lamenting the reactionary sexual marketplace which he sees as its consequence.[63] Houellebecq undertakes this critique by adopting aesthetically the features of the cultural and ideological atmosphere he decries. His fiction is replete with glorifications of consumerism, pornographic descriptions of sex, and misogynistic commentaries. This satirically cynical and yet ethical earnestness, Crowley and Best argue, 'juxtapose[s] irony and sincerity in a manner which makes it impossible to know which tone we should be taking seriously, if any'.[64] Houellebecq's work rebuffs any hermeneutic approach that seeks either to argue for its earnest satirical aims or its ironic questioning of the effectiveness of satire itself.

The figure of Daniel1 is an excellent example of the manner in which Houellebecq's work seems to resist attempts at a readerly command of its aims or sincerity. Daniel1 himself reflects on the ineffectiveness of his own grotesque, racist, violent, and misogynistic satire which, he observes, has been earnestly seized upon by politically progressive groups as a critique of the grotesque, racist, and violent world which they all share. Laughter, he begins to argue, is an expression of humanity's most deeply concealed cruelty, an expression of its willingness to delight openly in the very violence and sexism it claims to condemn. Moreover, laughter abjures any revolutionary possibility for satire; comedy, Daniel1 argues, does not 'transformer le monde' [transform the world], but makes it acceptable (*Possibilité*: 158; *Possibility*: 133–34). The satirist and his reader, Houellebecq argues through the voice of Daniel1, is in this respect a collaborator with evil: the satirist illuminates injustice while eradicating hope. So any critical attempt to assess the earnestness of Houellebecq's Utopian envisioning of an improved world, let alone his politically problematic critique of this world, is likely to be frustrated. This is not only because of the difficulty of establishing any ethical, ideological, generic ground in his work, but because the very act of doing so is anticipated and thus nullified by Houellebecq's proxy in the novel. In this way, Houellebecq through Daniel1 addresses his readers, condemning both their serious critical pretensions as well as their laughter. At the same Daniel1's acerbic, unforgiving picture of Western decadence anticipates the consequences of a world which has dispensed with critical circumspection and in which laughter is only available through ironic distance.

Readers of More's *Utopia* are divided between a commitment to reading it as a sincere and engaged political envisagement of a better world and seeing it as a humorous comment on England's venality. Reading each of these Utopian island novels too demands its readers to make a similar hermeneutic decision. Indeed,

63 Best and Crowley, *The New Pornographies*, p. 181.

64 Ibid. For other discussion of Houellebecq's use of irony see: Éric Bordas, 'Ironie de l'ironie', in *Les Lieux du réalisme: pour Philippe Hamon*, ed. by Vincent Jouve and Alain Pagès (L'Improviste, 2005); Clément Lemaitre, 'Mutations de l'ironie dans l'œuvre de Michel Houellebecq', *Carnets: Revue électronique d'études françaises de l'APEF*, 2nd ser., 23 (2022), doi:10.4000/carnets.13730.

where Zola's and Hardy's novels may be read as sincere responses to the problem of the Darwinian assault on humanity, and also to offer seriously considered — if not always successful — avenues of escape from the evolutionary real, it is not clear whether these works can be read as sincere responses to the Darwinian predicament, or indeed whether any putative re-conception of human being is offered in earnest. Yet satirical irony is not opposed to Utopian desire. In resisting and evading attempts to establish a hermeneutic foundation on which to ground interpretation, these novels themselves engage in Utopian acts of inscription and enclosure in the sense articulated by Marin. By attempting to protect themselves from a finalised or totalising interpretation, these novels re-enact a withdrawal of meaning that echoes the spatial coordinates of the Utopian fantasy, sealing each novel off as an impenetrable literary space. Indeed, if the ironic Utopian novel invites us to relinquish an unconscious Utopian desire for a 'correct' and exclusive reading, it at the same time instigates a yearning for closure by offering the tantalising prospect of its explicit representational and political ambiguity. Thus, if these novels seem to create a literary space in which a similarly sovereign (or Utopian) human might be articulated, it is a paradoxical one in which its imagined sovereignty is contingent on the impossibility of a final meaning.

Darwinian Contradiction and the Utopian Island

If the Utopian island is sovereign by virtue of its polysemy, Patrick Parrinder's work on the profusion of scientific Utopian literature in the nineteenth century underlines that Darwinian evolutionary utopianism has historically been more invested in a finality of meaning and being. Exploring Utopian fantasies of racial perfection in William Morris's *News from Nowhere* (1890) and Francis Galton's unpublished novel *Kantsaywhere*, Parrinder argues that a precondition of Utopian perfection is corporeal and physiological perfection. Utopian Darwinian literary writing, Parrinder suggests, seems less concerned with material or biological dynamism and openness, than it is with a social-Darwinian conception of the body politic and the individual body as sights of stability and closure. This desire, he argues, is derived not from More but from Plato's *Republic*, and that the discovery of the mechanics of Darwinian sexual selection in the nineteenth century enables us to conceive of enacting this perfectibility in material terms.[65] Eugenics (literally 'good breeding') as Mike Hawkins judiciously puts it, allow us to 'reduce [the] numbers [of undesirables] through relevant social controls — negative eugenics — while at the same time encouraging the reproduction of the better elements — positive eugenics'.[66] So-called 'positive eugenics', Parrinder suggests, amounts to a form of Utopian project, since it aims at the perfection of the human race as a prequel to the perfection of society. As such, the Utopian impulse is itself an integral part of evolutionary desire; sexual selection, he says, aims at 'good breeding' at

65 Patrick Parrinder, 'Eugenics and Utopia: Sexual Selection from Galton to Morris', *Utopian Studies*, 8.2 (1997), pp. 1–12.
66 Hawkins, *Social Darwinism in European and American Thought, 1860–1945*, p. 217.

an unconscious level.[67] Eugenics, therefore, may not be a Utopian perversion of evolutionary thought; rather, Darwinian evolutionary thought, it is implied, allows us to understand Utopian desire as biologically 'natural'.

Of course, such a reading, Grosz might argue, fails to grasp the dynamism of Darwinian sexual selection. Sexual selection is only partly (if at all) about reproduction and perfection, since the general economy of the sexual relation seems to problematise its own limited instrumental aims. Indeed, as I have emphasised, it is not clear that sexual selection is biologically necessary for reproduction in non-human organisms, suggesting that perfection is itself a human conceptual imposition on a non-human field of activity in which sexuality is but one of many strategies for reproduction and evolution. Hardy's work dramatises the collective, creative, and subversive nature of sexual relations, which in his work may be read to question the deterministic nature of evolutionary narrative. Yet, Parrinder's work demonstrates that the Utopian desire for perfection — for the creation of a human desert island — persists, and that such a desire, which he locates in the literary legacy of Darwin, may find a useful conceptual alliance in the teleological evolutionary narrative of development and progress (however accurate or coherent that is).

The apparently corporeally perfect inhabitants of Huxley's Pala seem oblivious to this conceptual tension in their own Utopian project, which involves a combination of religious mysticism, economic and social autarchy, and eugenic population policies, all in pursuit of the stability of their island enclave. Pala's claim is that its politically autarchic, economically isolationist polity owes its stability and serenity to structures of social organisation based on an application of Darwinian evolutionary theories, and the elevation of biological science to the status of a religious creed. Beginning at the end of the nineteenth century, the island's inhabitants, through a chance encounter with a Western evolutionist named McPhail, adopt alongside traditional religious practices a set of Darwinian principles applied to the task of personal and social improvement. This 'new wisdom', they say, is simply 'biological theory realized in living practice, is Darwinism raised to the level of compassion and spiritual insight' (*Island*: 193–94). Practically, this translates into eugenic breeding controls for humans and non-humans, enabling the genesis of 'a child of superior quality' (183) with selective reproductive planning while negating the necessity for external economic and political relations through the development of an efficient, self-sustaining agri-economy. These practices, combined with a faith in 'rational' Buddhism, allows the island and its inhabitants to remain biologically and politically separate from the world, and from its aggressively expansionist industrial neighbour, the Polynesian island of 'Rendang'.

If, as Krishnan Kumar claims, *Brave New World* was intended as a prescient warning against the enactment of Utopian desire through biopower, *Island* appears as its antithesis: an earnest, literary imagining of social improvement through Darwinian bio-politics.[68] The novel advances a reading of Darwin's thought,

67 Patrick Parrinder, *Utopian Literature and Science: From the Scientific Revolution to Brave New World and beyond* (Palgrave Macmillan, 2015), p. 13.
68 Krishan Kumar, *Utopianism* (University of Minnesota Press, 1991), p. 55.

specifically the application of sexual selection for 'positive' eugenics, not merely as a means by which biological transcendence is materially possible, but as the intellectual and religious lodestone around which society can organise itself. *Island*'s Darwinian programme for the perfection of society is based on making hegemonic a version of Darwin's thought that seeks to minimise the threat to humanity of evolution itself. However, just as Deleuze identifies the manner in which Defoe's desert island reveals the anti-materialist fallacy of its own implicit thesis, Huxley's island, despite itself, points to the paradox of its own conception of Darwinian evolution. As I have mentioned previously, Farnaby's chance entry into Pala is an explicit irruption of materiality into the island's space of transcendent autonomy, a pointed example of the way in which the island's isolation is always contingent. The contradiction in this conception is demonstrated by its educational ethos, as expounded by a Palanese schoolteacher:

> That's precisely the reason why we begin with [ecology]. Never give children a chance of imagining that anything exists in isolation. Make it plain from the very first that all living is a relationship. Show them relationships in the woods, in the fields, in the ponds and streams, in the village and country around it. Rub it in. (*Island*: 18)

While claiming to adopt a Darwinian ethos, the Palanese fail to recognise the anti-ecological and isolationist tendencies of their own Utopian pretensions.

Putting aside speculative accounts of Huxley's perceived intentions, however, the novel's didactic rhetoric can also be read despite itself as a warning against dogmatic Utopian thought and religious zeal. Through Pala and its hypocritical claims to an ethos of ecological relation we may recall Deleuze's diagnosis of the persistent psychic need for stability, according to which humans 'cannot live, nor live in security, unless they assume that the active struggle between earth and water is over, or at least contained'.[69] Despite Pala's self-imposed eugenic and political isolation, the novel concludes with Farnaby witnessing its imminent destruction in a military coup conducted by the heir to the island's throne, a young demagogue named Murugan. Just as Farnaby's infiltration of the country is an implicit repudiation of its pretensions to transcendent autonomy, Murugan's destruction of Pala is enabled by his escape from the insular culture of Pala. Visiting the neighbouring industrialist country of Rendang, Murugan is seduced by the material pleasures of capitalist consumption and expansion, triggering a desire to seize power in Pala. Huxley's novel, then, shows the persistence of the desert island fantasy in a Utopian context as well as the brutal consequences of its destruction. For all its totalising didacticism, *Island*, in this respect, can be read as a sobering warning against a fantasmic desire for human autonomy even as it impugns the violence of competitive human relationality.

Butler's *Erewhon* stages a similar set of conflicts as the novel's protagonist penetrates the nominally enclosed boundaries of the country of Erewhon, showing those boundaries to be porous and part of a dynamic material world. Just as Butler's depiction of the breaching of Erewhon's sealed space prefigures Huxley's similar

69 Deleuze, 'Desert Islands', p. 9.

gesture in the form of Farnaby's discovery of Pala, the conclusion of *Erewhon* anticipates the ending of *Island*. Having been imprisoned by the Erewhonians and forced to abide by their oppressive laws, the protagonist escapes, along with his Erewhonian lover Arowhena, in a hot air balloon, showing the borders of this Utopia to be both internally and externally porous. In its concluding pages, the novel's protagonist outlines plans for returning to Erewhon to colonise and exploit it as a source of labour. The narrator observes that despite its efforts to seal itself off from external life, 'one of the rivers which descends from the Snowy Mountains, and passes through Erewhon, is known to be navigable for several hundred miles from its mouth' (*Erewhon*: 255). He plans on penetrating Erewhon once again, this time with armed gunboats, to intimidate its inhabitants who have dispossessed themselves of all means of defence, and to coerce them into accepting wage-labour for a colonial sugar-growing company. The conclusion of Huxley's novel suggests that while sovereignty from evolutionary relation is impossible, a dynamic materialism may itself lead to violent conflict and forms of domination. Butler's work too stages a conflict between a political, social, and geographic autarchy founded on Darwinian principles and the irresistible violence of the survival of the fittest.

This dynamic is articulated particularly compellingly in *Erewhon*'s 'Book of the Machines', a document of the country's historical genesis, based on two earlier essays by Butler in which he speculates on the consequences of technological evolution. The novel's protagonist discovers on his arrival in Erewhon that the success of the country, as the Erewhonians see it, is based on their decision to destroy all machinery and prohibit the development of technology. Butler's narrator writes:

> I learnt that about four hundred years previously, the state of mechanical knowledge was far beyond our own, and was advancing with prodigious rapidity, until one of the most learned professors of hypothetics wrote an extraordinary book (from which I propose to give extracts later on), proving that the machines were ultimately destined to supplant the race of man, and to become instinct with a vitality as different from, and superior to, that of animals, as animal to vegetable life. (*Erehwon*: 97)

It is explained that the threat represented by technology is analogous to the threat represented by a rival species in the struggle to survive. According to the account in 'Book of the Machines', technology developed an evolutionary advantage over humanity, for where the human is mentally fallible and physically weak:

> The machine is brisk and active, when the man is weary; it is clear-headed and collected, when the man is stupid and dull; it needs no slumber, when man must sleep or drop; ever at its post, every ready for work, its alacrity never flags, its patience never gives in. (*Erewhon*: 205)

The destruction of machinery, ironically, represents a desperate attempt both to sustain human domination and, since dominance and autonomy is constitutive of humanity's self-conception, to reconstitute and sustain the very idea of humanity itself. The conclusion of Butler's novel, then, sharpens the contradictory nature of this selective understanding of evolution. By destroying machinery in the hopes of sustaining their own autonomy, the Erewhonians deprive themselves of any military technology or form of defence against potential external threats. The Erewhonians' assumption that their autonomy is secured by an internal intervention in evolutionary competition is shattered by the incursion from without of another violent antagonism.

A similar account of the contradictory and ultimately doomed nature of the Darwinian Utopian island unfolds in Houellebecq's *La Possibilité d'une île*. In his previous novel, *Les Particules élémentaires*, Houellebecq describes a type of neohuman life made possible by cloning, the non-relational mode of reproducing which means that for the human, in effect, evolution can be halted.[70] Similarly, in *La Possibilité d'une île*, since the neohuman does not age and reproduces only itself in an act of genetic repetition, each individual — and the species of neohumanity — becomes a biological desert island in an otherwise dynamic field of biological relations, hybridisations, and transformations. Like Butler's Erewhonians seeking to remove themselves from the world of evolutionary competition, the neohumans are a consequence of humanity's capacity to enact that desire for autonomy and put it into biological practice with genomic technology. Daniel25 comments on Nietzsche's characterisation of humanity as 'l'espèce dont le type n'est pas encore fixé' [the species whose type is not yet fixed], observing that 'si les humains ne justifiaient nullement une telle appréciation — moins en tout cas que la plupart des espèces animales — , elle ne s'applique pas davantage aux néo-humains qui prirent leur suite' [if humans in no way merited such an assessment — less so than most of the animal species in any case — it certainly no longer applies to the neohumans who followed them] (*Possibilité*: 413; *Possibility*: 361). Yet, the neohumans still depend on light, air, and minerals derived from their external environment, meaning that, like the Palanese, they are not strictly autonomous in relation to the external or material world. As with Butler's punctured Utopian enclave and Huxley's colonised island, Houellebecq's neohuman Utopia reveals the impossibility of the Utopian transcendence it appears to depict.

Furthermore, Houellebecq's novel interrogates the intolerable psychic toll of neohuman affective isolation. The life of the neohuman, for all that it offers serenity and freedom from the reproductive or competitive relation is a sterile form of emotional imprisonment. Yet, even this isolation is not fully rigorous: human relations do manage to persist through the neohumans' exegetical commentary on the life stories of their genetic predecessors, in dialogue with whom the neohumans come to realise the social and relational vacuum in which they have come to live. In digital conversation with another neohuman, Daniel25 discovers that one of his former correspondents, Marie23, has been driven to escape her own compound, a decision that effectively ends her life. She undertakes it with the intention of seeking out human relations in the external world in response to the ecstatic suicide note of Daniel1 in which he affirms the redemptive, timeless nature of love. Reflecting upon this decision, Daniel25's interlocutor observes that:

> Cette lettre qui a eu sur Marie23 une influence catastrophique, qui l'a poussée

70 McCann, *Michel Houellebecq*, p. 174.

> à partir, à s'imaginer qu'une communauté sociale — d'humains ou de néo-humains, au fond elle ne savait pas très bien — s'était formée quelque part, et qu'elle avait découvert un nouveau mode d'organisation relationnelle; que la séparation individuelle radicale que nous connaissons pouvait être abolie dès maintenant, sans attendre l'avènement des Futurs. (*Possibilité*: 432)
>
> [This letter that has had a catastrophic effect on Marie23, and drove her to leave, to imagine that a social community — of humans and neohumans, basically she didn't really know — had formed somewhere, and that she had discovered a new mode of relational organization; that the radical individual separation we now know could be abolished immediately.] (*Possibility*: 376–77)

However, just as Marie23 is compelled to abolish isolation and vacate her compound in search of affective and social connection, Daniel25 is finally stimulated to do the same after reading Daniel1s' final poem addressed to his lover, Esther. Erewhon's technological application of the Darwinian principle of competition turns out also to be its undoing, as its destruction of technology leaves the country vulnerable to enemies; the Palanese delusion of transcendence from socio-economic relations, based on a Darwinian biopolitics of technology, blinds them to material threats to their Utopia. In Houellebecq's neohuman Utopia, the techniques of reading, re-reading, and commentary which enable neohumanity to transcend evolution, are also the very conditions which destroy that transcendence, drawing the neohumans back to social and biological materiality, the promise of relational life, and death.

Each of these novels, then, demonstrates the fantasmic and mythic nature of a timeless, autarchic, socio-political and biological territory, the evolutionary Utopian island. The desert island is an ideation whose fantastical and impossible nature is confirmed by the laws of evolutionary naturalism in which its creators place their faith. As such, we may see how these Utopian novels appear paradoxically to reproduce the putative determinism and pessimism of post-Darwinian literary realism, where biological reality is thought to negate the prospect of liberation. Commenting on the specific case of Houellebecq, Crowley and Best make the observation that this type of self-defeating Utopianism is a particularly despairing literary form:

> When utopia is no longer available as the end point of an ultimately eschatological progressive narrative [...] it can only be configured as an attempt at withdrawal, a sidestep, an interruption, a puncture — a kind of utopia under erasure. [...]
>
> But if utopia is only conceivable as a puncture, or as under erasure, then any denunciation of all the misery that surrounds this utopia may well find itself also punctured, semi-effaced, irretrievably in hock to this hegemonic misery.[71]

In Houellebecq, as in Huxley and Butler, neither the post-Darwinian Utopian fantasy nor its satirical denunciation can be redemptive. Both the earnest act of imagining an improved future and the critical act of condemnation work in the service of prevailing materialist hopelessness.

71 Best and Crowley, *The New Pornographies*, pp. 209–10.

We may, however, read the relation between Utopia and satire in a different way. Socio-political and biological forms of Utopian withdrawal are shown in Houellebecq's, Huxley's, and Butler's stories to be symptomatic of an unrealisable desire for timeless sovereignty. Yet, as I indicated in this book's introductory discussion of Derrida's reflection on the animal, such a desire — both by being a desire and by being impossible — may be interpreted in affirmative terms. These Utopian novels, ironically, are mimetically faithful to the evolutionary narrative of human life, insofar as they dramatise and formally instantiate a desire to exceed evolutionary materiality via the creation of a sovereign human and, in puncturing that desire, seem to acknowledge not only its impossibility but its undesirability. Humanity in these works is a temporary, contingent, and porous Utopia, whose value need not lie in how it concludes (for it will always end) but in the possibilities and relations it proliferates while it exists.

The Dialectics of Darwinian Utopias

Louis Marin shares this affirmative understanding of island creation; the Utopian act of withdrawal is a positive act of inscription at the same time as it is a violent incision. Abraxas is destroyed when a monumental trench is dug and it becomes cut off from the continent to which it was physically attached, the condition by which Utopia is created. However, Marin points out, Utopia can never truly become independent of relation and connection, and instead seeks to bury its connections to the continent from which it came by bribing its neighbours, quietly assassinating its enemies, and repudiating international treaties.[72] This act of disingenuous insulation is mirrored by More who writes *Utopia* in the form of a factual travel narrative, creating a new form of discourse by disguising its fictional origins. Yet, just as More's work describes the act of incision that creates the island, thereby undermining the island's own attempt to conceal its origins, Hythloday's description of the island of Utopia and its uncanny resemblance to England, its phantom referent, undermines More's assertion of the text's facticity. Utopia is similar in size to England, although different in shape; the number of Utopian cities, fifty-three, matches the number of counties in England and Wales in the sixteenth century.[73] For all that the country of Utopia appears historically, geographically, and politically disconnected, as Christopher Kendrick points out, England 'provides [Utopia] with the raw material from which [the text] is spun'.[74] Utopia therefore is enchained to its source material — the earth, history, political reality — by a variety of different forms of relation. For at the same time as Utopia represents a negation of its referent, it also narrates the process by which Utopia seeks to efface the trace of that referent, revealing Utopia in its representation by More to be a negation of, a mirror of, and penetrated by its origins all at the same time.

72 Marin, *Utopics*, pp. 99–100.

73 Logan and Adams refer to William Harrison's 1587 *Description of England* in which this tally is made: More, *Utopia*, p. 43, n. 6.

74 Christopher Kendrick, 'More's Utopia and Uneven Development', *Boundary 2*, 13.2–3 (1985), pp. 233–66 (p. 243).

For Marin *Utopia* represents a 'textual staging' of the contradictions that constitute textual and discursive creation. Utopia is a story about discursive foundation, describing the conditions of possibility that make texts — and the novel, *Utopia* — possible, addressing the act of concealment which grants More's text political urgency. The creation of the island does not simply seek to create a desert island and fail, but 'engenders a plurality of spaces in the totality of one project'.[75] The violent cutting away of relation with the intention of creating a political, spatial, or discursive totality proliferates relations between discursive, political, geographical, and historical spaces, formalising and staging these relations in a single text. This formal point about the creation of Utopian textual space has consequences for the way we read the treatment of Darwinian evolution in *Erewhon*, *Island*, and *La Possibilité d'une île*, each of which stages an attempt by humanity to withdraw and cut itself off from history, biology, and social relations, through the creation of an island. While each of these works narrates the failure of these attempts, we cannot conclude simply that islands of this sort are materially impossible. Rather, the ephemeral and precarious nature of Erewhon, Pala, and the neohuman state of Daniel24 and Daniel25, underscores the nature of their historical, social, and biological materiality which — like the individual organism — is both their condition of possibility and the cause of their failure. Rather than fixate on their failure, we might look instead to the relations and thought that these literary islands enable while extant as we look further into the internal conflicts they illuminate between the positive Utopian project of human transcendence and the evolutionary predicament it seeks to escape.

Huxley's *Island* reveals the contradictions in Pala's attempt to engender an autarchic state through the adoption of Darwinian eugenics and the creation of a scientific religion. The final scene of the novel depicts the ultimate consequences of this contradiction, as the form of the Darwinian, competitive imperialism of Rendang, of which Pala is intended to be a definitive negation, undoes its Utopian autarchy from within, in the form of Murugan. However, the novel's depiction of Utopian self-enclosure can be read as more than a reconfirmation of the irreducible antagonism between a nihilistic acceptance and political enactment of the Darwinian 'war of nature' against Utopian naïvety. Rather, this conflict allows us to see the continuities between these two iterations of Darwinism: the laissez-faire nihilism of the war of nature is implicitly Utopian at the same time as it is self-destructive. As one educated Palanese man points out, Rendang's population continues to increase at an unsustainably high rate, a paradoxical success of unchecked capitalist growth since it results in widespread poverty and misery, threatening the very stability of the country. Such a position demands what Marxist geographers call the 'spatial fix': as the demand for growth exhausts the resources and labour of one's own territory, imperial expansion is required to maintain existing levels of profit.[76] So even as Rendang's embrace of evolutionary antagonism implies a certain acceptance of the

75 Marin, *Utopics*, p. 12.

76 David Harvey, 'Globalization and the "Spatial Fix"', *Geographische Revue: Zeitschrift für Literatur und Diskussion*, 3.2 (2001), pp. 23–30.

inherent violence of evolutionary competition, its project of constant economic growth is conducted to sustain a dominance at all costs, even if constant, unchecked growth requires Malthusian elimination or reaches its logical conclusion in collapse. By contrast with Pala's rigid eugenic programme for population control which demands an impossible level of biological control, Rendang's dogmatic anarcho-liberalism reveals the illusions of immortality concealed by heedless consumption and nihilism.

Pala's Utopian desire for autarchy seeks to efface any historical trace of this same nihilism within its own body politic, but finds that trace revealing itself in the figure of Murugan. But Murugan understands his action in precisely the opposite way, arguing that imperialism and capitalism are the logical conclusion of the history of Pala and its embrace of Darwinian evolution. He argues that his plans for militarisation, industrialisation, and opening Pala's borders to trade is a continuation of 'the revolution that was started more than a hundred years ago' (*Island*: 47) by the evolutionary scientist McPhail, who had introduced Western science to Pala. Murugan, then, reintroduces the spectre of the history of evolutionary theory, specifically the nineteenth-century origins of evolutionary Darwinism, and implicitly alludes to the ideological climate of imperialism and capitalist industrialism of which Darwinian natural selection was a reflection and which it sustained in turn. For all that Pala seeks to repudiate ideology and history through its Utopian reading of Darwinian evolution, it is the persistence of Darwinism's ideological undertow which leads to the destruction of Pala.

Butler's *Erewhon* also stages a dialectically interlaced set of relations between ostensibly divergent iterations of Darwinism. Erewhon symbolises both an inversion and mirror image of Victorian Britain: the initially strange and corrupt world of Erewhon comes to reflect the injustices of Victorian society. Each of these worlds, the imaginary construction of Erewhon and its implicit referent, Butler's Victorian milieu, corresponds to nominally contrary interpretations of Darwinian evolution, specifically in relation to technology. 'The Book of the Machines' espouses the destruction of all machinery, arguing that technology represents an evolutionary competitor to humanity. By contrast, Darwinian natural selection seems to complement the Victorian laissez-faire economics of free-market competition which, Herbert Sussman argues, nourished the ideological climate of the industrial revolution.[77] Read in terms of Herbert Spencer's concept of evolution as a movement from simplicity to complex, technological progress represents but one of the drivers and outcomes of a form of evolutionary progress that has humanity as its telos. Yet, these contrasting visions of evolution have at their core an identical goal: to sustain the dominance of the human.

Marx's understanding of the Luddite movement allows us to develop this relationship between a (specifically Victorian) embrace of technological advancement and evolutionary progress on the one hand, and, on the other hand, Erewhon's Darwinian justifications for the wholesale rejection of technology. In the first

77 Herbert L. Sussman, *Victorian Technology: Invention, Innovation, and the Rise of the Machine* (Praeger, 2009), p. 3.

volume of *Das Kapital* (1867), Marx identifies the struggle between technology and wage-labour in terms of the fundamental contradictions inherent to capitalism, outlining the ways in which the increasing speed of mechanisation renders skilled labourers obsolete, stripping them of economic agency.[78] Butler's Erewhon might be read as a desire to withdraw from such exploitative economic relations, and metonymically from the Darwinian and economic war of nature, which, in the context of the Industrial Revolution in Britain and its afterlife in the nineteenth century, can be exemplified by the struggle between the human and the inhuman machine. Marx's response to the exploitative nature of technological evolution, however, is not destructive. The Luddites, he argues, for whom the Erewhonians are a plausible literary cipher, were mistaken in identifying technological evolution as the source of their exploitation. They should instead have 'directed their attacks, not against the material instruments of production, but against the mode in which they are used'.[79] That is to say, Marx advocates for an analysis of the relations between technology and humanity, illuminating the specific ways in which these relations allow certain 'modes' of exploitative relation, rather than attacking the objects as such.

This focus on the relations between labour and technology anticipates Deleuze and Guattari's argument in *Anti-Oedipus* that the ontological difference between the human and the machine is insubstantial. Drawing on Marx's *Economic and Philosophic Manuscripts of 1844*, Deleuze and Guattari cite his brief allusion to the dynamics of human sexuality as a radical intervention in theorising human-machine relations.[80] Where Marx writes that 'the direct, natural, and necessary relation of person to person' is that of the relation between man and woman, Deleuze and Guattari argue that this is a reflection on the 'molarity' of specifically human sexuality amidst myriad forms of other 'molecular' libidinal relations that exceed the oppressive network of the Oedipal family.[81] Similarly to Grosz, who argues that the libidinal relations between individual organisms should be understood within the context of a co-creative dynamic between culture and nature, Deleuze and Guattari argue that a merely biologically reproductive understanding of the sexual relation of person to person lacks the dynamic possibilities engendered by non-human forms of libidinal relations. Human relations with the non-human catalyse an explosive increase in reproductive potentialities, opening up the possibility of new life-forms, or dynamic assemblages, that are constituted by multiple, interrelated beings, both human, non-human, and even virtual.

78 Karl Marx, *Capital. Volume One: A Critique of Political Economy*, ed. by Friedrich Engels, trans. by Samuel Moore and Edward Aveling (Dover, 2012), pp. 466–71.

79 Ibid., p. 468.

80 Gilles Deleuze and Félix Guattari, *Anti-Oedipus: Capitalism and Schizophrenia*, trans. by Robert Hurley, Mark Seem and Helen R. Lane (University of Minnesota Press, 1983), p. 294.

81 Karl Marx, *Economic and Philosophic Manuscripts of 1844*, ed. and trans. by Martin Milligan (Dover, 2012), p. 101.

Further analysis of 'The Book of the Machines' reveals a similar conception of the relation between humanity and technology as outlined by the Erewhonians. *Erewhon*'s narrator offers the following excerpt:

> 'How greatly,' he wrote, 'do we not now live with our external limbs? We vary our physique with the seasons, with age, with advancing or decreasing wealth. If it is wet we are furnished with an organ commonly called an umbrella, and which is designed for the purpose of protecting our clothes or our skins from the injurious effects of rain. Man has now many extra-corporeal members, which are of more importance to him than a good deal of his hair, or at any rate than his whiskers'. (*Erewhon*: 224)

Humanity's dependence on technological prosthesis, he argues, has become so thorough that technology must cease to be viewed as wholly organic, but as an assemblage of human and non-human organs. In fact, the fictional author of this work argues, such is the deep evolutionary history of technological prosthesis that the very idea of the human itself comes under question, 'the earliest accidental use of the stick having set the ball rolling' (*Erewhon*: 223). This mirrors Friedrich Engels's reading of Darwin's theory of the law of 'correlation of growth' in 'The Part Played by Labour in the Transition from Ape to Man'. Labour and the making of tools, Engels argues, inaugurates humanity and its distinction from animality, leading to the paradoxical conclusion that the human begins with the incorporation of non-human organs: the point when humanity as a category ends.[82] The human island, according to this reading, is neither a non-existent entity lost in evolutionary flux, nor a transcendent category infused with supreme agency. The human in these accounts exists as a physical entity whose capacity to exploit the supplementary quality of its external environment makes a definitive image of that entity impossible.

Butler, therefore, stages a dialectic between a Victorian belief in technological advancement as a correlative of evolutionary progress and the rejection of that advancement by the Erewhonians. This in turn allows us to see how both Erewhon and the Victorian progressive mind-set share a common Utopian aim, the maintenance of human biological and ontological supremacy. Butler's Erewhonians do not merely fear being made extinct by technology, but more fundamentally fear being rendered part of a larger technological reproductive assemblage, which implies that the human is not inherently autonomous but that it already exists as part of a network of bio-technological relations. Conversely, even though the Victorian emphasis on technological evolution is aimed at human advancement, the result of this evolution is, as Marx, Deleuze, and Butler suggest, the erasure of the very human autonomy upon which humanity's self-conception depends. In that respect, the Luddism of the Erewhonians is merely an intensification and, in a way, a more explicit version of the egotism that drives a faith in progress with humanity as its telos.

82 Friedrich Engels, 'The Part Played by Labour in the Transition from Ape to Man' in *Dialectics of Nature*, trans. by Clemens Dutt, 2nd rev. edn (Progress, 1986), pp. 170–83 (p. 172).

Butler's negative image of the co-constitutive nature of human and non-human anticipates with remarkable prescience Donna Haraway's *A Cyborg Manifesto* on the feminist possibilities engendered by the relation between humanity and technology.[83] Grosz, as I have shown, offers an affirmative reading (if limited by its own tendency to reify sexual dimorphism) of the libidinal relations between the sexes which argues that the relations encoded in biological sexuality engender a set of possibilities that exceed that initial relation. Haraway posits an affirmative interpretation of the possibilities of human-machine relations which is premised on the biological indeterminacy of humanity, arguing that new digital technologies open up the possibility of an indeterminate 'cyborg' ontology through which it is possible to write and rewrite social, political, economic, and gender realities. Becoming integrated with technology, becoming inhuman, she says, might not be dreaded as it is for the Erewhonians, but should be embraced and harnessed 'through the skilful task of reconstructing the boundaries of daily life, in partial connection with others, in communication with all of our parts'.[84] Haraway's theory of cyborg indeterminacy celebrates the dissolution of the category of the transcendent, rational, and sovereign human, a category, Freud observed, to which we as a species resiliently hold fast, despite (or perhaps because of) the advent of the Darwinian revolution.

The cyborg in *La Possibilité d'une île* is tasked with precisely the opposite goal. The neohuman is an effort at total evolutionary, biological stasis — an attempt at a determinism so complete that its outcome is the absolute negation of change. And although we know that this effort is ultimately unsuccessful, in creating that temporary textual and conceptual island space, Houellebecq engenders within a single text a set of relations between two apparently divergent bio-political Darwinisms. On the one hand, the narrative of Daniel1 is dedicated to describing the conditions of a contemporary Western society where what Sylvia Wynter calls Darwin's 'redescription of the human' has resulted in an intensification of the capitalist ideology of ruthless competition as natural law.[85] On the other hand, the Utopian state of neohumanity described in the narratives of Daniel24 and 25 is designed to escape and negate that world by instantiating a state of being where the advancement of Darwinian science and the development of genomic technology has engendered a world devoid of competition and human relation as such. Where More's *Utopia*, Huxley's *Island*, and Butler's *Erewhon* create implicitly dialectical relations between the Utopian state of withdrawal and its phantom referent, Houellebecq makes these relations explicit by formally intertwining the three narratives of Daniel1, 24, and 25, constructing a literary structure analogous to the double helix model of DNA.[86]

83 Donna Haraway, 'A Cyborg Manifesto: Science, Technology, and Socialist-Feminism in the Late Twentieth Century', in *Simians, Cyborgs and Women: The Reinvention of Nature* (Routledge, 1991), pp. 149–81.

84 Ibid., p. 456.

85 Sylvia Wynter, 'Unsettling the Coloniality of Being/ Power/ Truth/ Freedom: Towards the Human, after Man, Its Overrepresentation — An Argument', *CR: The New Centennial Review*, 3.3 (2003), pp. 257–337 (p. 308), doi:10.1353/ncr.2004.0015.

86 McCann, *Michel Houellebecq*, p. 174.

Douglas Morrey observes that in DanielI's twenty-first-century world of neoliberal, competitive hegemony, the economic urge for domination is understood to be driven by an evolutionary compulsion for survival. The cruelty of applying free-market logic to sexual relations, Morrey argues, is not merely a function of the extension of capitalist rationality to libidinal relations; it originates in 'a deep atavistic heritage that serves to underline these aspects of human behaviour as so many evolutionary facts'.[87] This line of reasoning is borne out by the conception of human sexual competitiveness that DanielI's first wife, Isabelle, identifies as originating in a primitive instinctual will to power. Explaining her success as an editor of magazines for teenage girls she elaborates upon this theory:

> Si les filles sont attirées sexuellement par les types qui montent sur scène, poursuivit-elle, [...] c'est aussi qu'elles sentent qu'un individu qui monte sur scène risque sa peau, parce que le public est un gros animal dangereux, et qu'il peut à tout instant anéantir sa créature, la chasser, l'obliger à s'enfuir sous la honte et les quolibets. La récompense qu'elles peuvent offrir au type qui risque sa peau en montant sur scène, c'est leur corps; c'est exactement la même chose qu'avec un gladiateur, ou un torero. Il serait stupide de s'imaginer que ces mécanismes primitifs ont disparu: je les connais, je les utilise, je gagne ma vie avec. (*Possibilité*: 36)
>
> [If girls are sexually attracted to guys who get up on stage [...] it's not simply that they are seeking fame; it's also that they feel an individual who gets up on stage risks his neck, because the public is a big dangerous animal that can annihilate its creation, hunt it down, and force it to flee, booed off in shame. The reward these girls can offer to the guy who risks his neck by going on stage is their body; it's exactly the same thing with a gladiator, or a matador. It would be stupid to imagine that these primitive mechanisms have disappeared; I know them, I use them, I earn my living from them.] (*Possibility*: 25)

Eliding the restrictive economy of natural selection with the general economy of sexual relation, Isabelle reduces sexual desire to a market-driven competition, in which the failure of an individual is be understood to be fatal. DanielI both benefits and suffers from this competitive, individualist conception of human relation and sexuality. He uses his wealth and status to secure sexual gratification while also succumbing to the cruelty of a system that organises its social and sexual hierarchy along biologically essentialist lines which sidelines him as he ages. Consequently, DanielI's relationship with the much younger actress, Esther, is both a source of profound pleasure and existential pain. With Esther, Daniel loses himself in the ecstasy of orgasm and revels in the enactment of the pornographic fantasies reified by a society organised around heteronormative and rigidly hierarchical conceptions of sexuality; at the same time he is constantly aware of his age and his superfluity to Esther, who herself is an exemplary sexual consumer. It is partly for this reason, we are to assume, that the relational and evolutionary stasis of neohumanity is attractive to DanielI, since it offers a renunciation of all forms of relation — those that cause pleasure as well as those that cause pain.

87 Douglas Morrey, *Michel Houellebecq: Humanity and Its Aftermath* (Liverpool University Press, 2013), p. 131.

Reflecting upon the pain which humanity's evolutionary position entails, Daniel24 outlines the reasons that neohumanity represents such a welcome proposition:

> L'homme avait un cerveau de grande taille, un cerveau disproportionné par rapport aux exigences primitives engendrées par le maintien de la survie, par la quête élémentaire de la nourriture et du sexe; nous allions enfin pouvoir commencer à l'utiliser. Aucune culture de l'esprit, me rappela-t-il, n'avait jamais pu se développer dans les sociétés à délinquance forte, simplement parce que la sécurité physique est la condition de la pensée libre, qu'aucune réflexion, aucune poésie, aucune pensée un tant soi peu créative n'a jamais pu naître chez un individu qui doit se préoccuper de sa survie, qui doit être constamment *sur ses gardes*. (*Possibilité*: 324–25)
>
> [Man had a large brain, disproportionate in relation to the primitive demands arising from the struggle to survive, from the elementary quest for food and sex; we were, at last, going to be able to use it. No culture of the mind, he reminded me, had ever been able to develop in societies with a high level of delinquency, simply because physical security is the condition for free thought; no reflection, no poetry nor idea of the slightest creativity has ever been engendered in an individual who has to worry about his survival, who has to be constantly on his guard.] (*Possibility*: 280)

Here, Daniel24 reaffirms the idea that the very evolutionary theories that consign us to materialist misery can also deliver us from that predicament. In that way, Houellebecq shows the neohuman island not as a break with an archaic, mortal humanity, but as a logical development of that humanity's desire for transcendence. More than that, Daniel24 suggests, by freeing itself from the struggle to survive, neohumanity ironically echoes the very individualism that causes so much misery in Daniel1's own time. The image of the neohuman enacts the Utopian work of rediscovering and reaffirming the transcendental figure of the human individual, pushing the image of sovereign individual to its logical end, seeking to become immune to the very conditions that produced it.

In these works, the Utopian island is not a transcendent figure, but a textual, material, and dynamic entity whose function it is to encode a multiplicity of relations in a single text. In this way, the Utopic act of self-enclosure, formalised by the hermeneutic repudiation provided by satirical irony, is undertaken with the intention of proliferating relations: between readings of Darwinisms, between visions of humanity; and between the Utopian envisagement of futurity and the trace of reality from which it is derived. This proliferation of relations allows us to trace the way in which various readings of Darwin are mobilised in the service either of Utopian idealism or desperate nihilism, showing the relationship between the two to be one of continuity. The withdrawal of life involved in Utopia is itself a form of nihilistic renunciation, while the hedonistic embrace of evolutionary competition is itself Utopian in its underlying assumption of immunity from the consequences of that violence.

Conclusion: 'closing brackets on becoming'

In Houellebecq's *La Possibilité d'une île*, Daniel25 describes to his future readers the rationale behind the neohuman, Utopian impulse. His explanation is not couched in biological or evolutionary terms, but metaphysical ones: 'Rejetant le paradigme incomplet de la forme, nous aspirons à rejoindre l'univers des potentialités innombrables. Refermant la parenthèse du devenir, nous sommes dès à présent entrés dans un état de stase illimité, indéfini' [Rejecting the incomplete paradigm of form, we aspire to rejoin the universe of countless potentialities. Closing the brackets on becoming, we are from now on in unlimited, indefinite stasis] (*Possibilité*: 426; *Possibility*: 372). This justification suggests an interaction between the ontological, the biological, and Utopian desire, the relations between which in three Utopian novels have been the focus of this chapter. For the neohumans, the Utopian desire for ontological stasis — a state of pure, timeless, metaphysical being — is realisable through the applied laws of evolution and genetics. But in this the contradictory nature of the neohumans' Utopian project is also disclosed. The neohumans reject the incompletion of form but embrace evolutionary thought, which, as Deleuze and Grosz show, reveals the durational nature of biological materiality, the constant becoming of natural life.

This relationship between a desire for impossible ontological and biological stasis and the attempt at realising that desire with applied versions of Darwinian evolutionary thought runs throughout the novels I have read in this chapter. Erewhon's rejection of becoming a cog in a dynamic, machinic assemblage is the expression of a desire to secure biological autonomy for the human from evolutionary change, stasis for the human being. Yet in rejecting incompletion through the destruction of machinery, the illusory nature of their ontological, societal, and evolutionary isolation is revealed with force, as their lack of machinery leads to the downfall of their Utopian enclave. Similarly, the Palanese, for all that they espouse ecological interconnection, desire more than anything autonomy from any political, social, and evolutionary network of relation and reproduction that threatens the racial and ideological purity of their Utopian settlement. Like the Erewhonites, the illusory nature of the Palanese Utopian isolation is revealed to them with brute force, as the socio-political aspect of Darwin's war of nature reasserts itself in the shape of Rendang.

The spatial figure of the island and the dialectical form of the Utopian novel allows these novels to articulate these contradictions. In the process, they allow us once more to explore the relationship between nihilistic and optimistic, transcendent responses to Darwin's theory of evolution. While showing the Utopian islands to be guilty of what Deleuze in his essay on 'Desert Islands' reveals as transcendent delusions, each of these novels is equally concerned to show that anti-Utopian nihilism is as delusional as it is dangerous. This has significant consequences for our understanding of the historical and social milieus of which each of these novels, with a combination of Utopian affirmation as well as satirical negation, is offering a critique. In *Erewhon*, the reactionary Utopianism of Luddism is shown as an extension of the nihilism of industrial capitalism; in *Island*, the spiritual, post-ideological delusions of Pala and its programme of human eugenics is revealed to be historically and ideologically reliant on the nineteenth-century industrial capitalism it decries; and in *The Possibility of an Island*, the supposedly transcendent, anti-relational genetic Utopia is exposed as an intensification of the cruel and atomised capitalist individualism, which the invention of the neohuman was designed to escape. Nihilist and Utopian responses to Darwin's assertion of humanity's animality are shown equally to be aimed at a desperate reconstitution of the human.

With that said, each of the novels I have discussed here acknowledges the persistence of Utopian desire, is sympathetic to that desire, and reminds us both of the productivity of its imaginary consequences as well as its potential dangers. They show how Utopian withdrawal can offer humanity the experience of pleasure, the prospect of redemption and of real agency, even as these are granted to some at the exclusion of others and are ultimately ephemeral. But in those temporary Utopian enclaves there is a trace of the dead reality from which they came, a remainder of what the imaginary desert island seeks to repudiate.

The Utopian island in these works gives literary form to the combined sense of agency and brutal limitation that characterises human life in the light of Darwin. The human, like the island, is a contained dynamism: both real and always on the verge of death. These works give voice to that reality — to the peace of isolation, to the enjoyment of our Utopian delusions — as well as to the necessary telos of all evolutionary life, death. As reflections on the general condition of the human as an evolutionary being, these works remind us that the human is an ontological impossibility but a performative reality, made possible paradoxically by humanity's inscription of itself into the world.

CONCLUSION

❖

Reading Darwin: Paranoia and Reparation

This book begins by asking: how does humanity respond to the egoic, epistemological, and ontological death sentence pronounced by Darwinian evolutionary thought? Darwin's theory of evolution shows that humanity is united with and derived from the animal and puts paid to any notion of anthropocentric autonomy. On the other hand, such a destruction of human sovereignty also problematises the epistemological authority upon which Darwin's theory subsists as well as the ontological status of the human as such. By revealing the human to be biologically indissociable from its object of inquiry — 'nature', 'the animal' — Darwin questions the epistemological universality implicit in his own theory and its moral ontology: evolutionary thought is revealed to be symptomatic of the same anthropocentrism it asks us to reject.

This book has sought to address this tension in Darwin's theory of evolution by looking specifically at responses to it in literary art. I have sought to demonstrate that the very fact of literature's existence and its capacity to respond to Darwin's evolutionary thought presupposes a kind of unique human agency which complicates humanity's inaugural fall from grace with which the argument begins. Darwin destroys humanity as a transcendent category, but does not destroy humanity itself. Humanity instead can be understood as performative, inscribing itself into the world, not definitively, but in a way that is contingent on that act of inscription itself. Similarly, we can read the Cartesian epistemological position of scientific autonomy that Darwin assumes as a kind of performance, a staging of authority that allows certain thoughts to be articulated, even if those thoughts later undercut their epistemological condition of possibility. I have sought, therefore, to explore how the performative truth of literary art responds to Darwin's theory of evolution; specifically, how in their responses to Darwin the novels I have studied themselves perform complex, contradictory readings of Darwin's theory of evolution, and seek to reconstitute a new form of human in the wake of the Darwinian revolution.

What modes of reading is it possible to employ, when the biological, ontological, and epistemological positions from which to read are unstable? Examining the contemporary usage of what Paul Ricœur calls the 'hermeneutics of suspicion', Eve Kosofky Sedgwick suggests that two interpretive *ethea*, two modes of reading, are available to the critical subject: 'paranoid' reading and 'reparative' reading.[1] Paranoid reading, Sedgwick suggests, has its origin in the works of the 'masters of suspicion', Freud, Marx, and Nietzsche. Their 'destructive' critical methods privilege exposure; seek to anticipate surprise and thus to negate its existence; and endeavour at all times to apply themselves to the broadest and most diverse spectrum of phenomena possible by implicitly tracing their emergence to a single source. Reparative reading, by contrast, is less concerned with the exposure of false consciousness, with anticipation, or with wide applicability. It seeks instead to explore the specific contingencies and local consequences that arise out of the performative nature of knowledge. In other words, by insisting on the need 'to hypothetically disentangle the question of truth value from the question of performative effect', reparative reading attempts to respond to a given body of knowledge or theory by exploring how best to 'move amongst its causes and effects'.[2] Without accepting that a truth is given, the reparative reader nevertheless seeks ways to relate in an affirmative way and without self-destructive paranoia in relation to that truth. Even true paranoiacs, Sedgwick argues, have real enemies: the paranoid ethos can never be paranoid enough, seeking as it does to reconfirm interminably its originary neurosis. Reparation is dedicated to the epistemological primacy of the affective life of humanity, to the recovery of that life, and seeks to resist the categorical imperative associated with paranoia which, in Sedgwick's view, has obscured the otherwise productively diagnostic practice of critical suspicion. Reparation is committed to negotiating truth as it is constructed, and finding redemptive possibility and consolation in that process of construction.

Sedgwick's notion of reparative reading is a productive one when it comes to assessing critically the work of scientific naturalism. Alluding to the conspiracy theories that plagued the HIV crisis in the last two decades of the twentieth century, Sedgwick reflects upon the value of knowing the 'probable natural history of HIV'. Suppose we knew that HIV originated as a deliberately engineered virus designed by the USA to decimate non-combatant enemies, she asks, rather than a biologically contingent and unpredictable viral emergence? What specific epistemological and practical consequences would that have? Suppose science could reveal definitively that conspiracy to be true, exposing the state's indifference to the deaths of millions in developing countries: 'what would we know that we don't already know?'[3]

Sedgwick here challenges the idea of the intrinsic value of truth, versus the question of what a truth makes possible. The question suggests that the epistemological ethos of natural history by definition engenders a paranoid method of reading life. The categorical imperative of orthodox institutional science is often to reveal truths, demystify false assumptions, reduce as many diverse phenomena to as many little causes as possible, and to eliminate surprise from humanity's view of

1 Eve Kosofsky Sedgwick, 'Paranoid Reading, Reparative Reading, or You're so Paranoid, You Probably Think This Essay Is about You', in *Touching Feeling: Affect, Pedagogy, Performativity* (Duke University Press, 2003), pp. 123–51.

2 Ibid., p. 129. Paul Ricœur, *Freud and Philosophy: An Essay on Interpretation*, trans. by Denis Savage (Yale University Press, 1970) pp. 32–33.

3 Sedgwick, 'Paranoid Reading', pp. 123–24.

nature. Darwin's theory of evolution can be viewed in the same way. And like those two other great demystifying figures in Western scientific thought, Copernicus and Freud, Darwin's theory of evolution exposes what, for Grosz, is the fundamental element of all false consciousness: humanity's assertion of ontological sovereignty. On the other hand, Sedgwick also asks us to consider how we as critical readers can relate to that destructive gesture. What if the Darwinian conspiracy is true and, as Daniel1 puts in *La Possibilité d'une île*, human life is nothing but 'un joli arrangement de particules, une surface lisse, sans individualité, dont la disparition n'aurait aucune importance' [a pretty arrangement of particles, a smooth surface, without individuality, whose disappearance would hold no importance] (*Possibilité*, 338; *Possibility*, 292); a meaningless, valueless, biological contingency, guilty of deluding itself of the opposite by virtue of the human capacity to think? What would we then know that we did not already? Certainly, Darwin's work can be understood as adding further scientific credence to an already existing economic, philosophical, and naturalist pessimism outlined by the likes of Hobbes, Schopenhauer, and Malthus, which seems to be extended in Zola, Hardy, and the collapse of a series of Darwinian Utopias. A more productive question might be then: in what way can we relate to the pessimistic truth of Darwin's theory of evolution as it is constructed without recapitulating the paranoia this truth seems to beget? If it is true that the human has no categorical biological existence or ontological fixity, what can be recovered in the thinking of human life nevertheless? It is this question which has driven my reading of responses to Darwin in novel form. It is also a persistent question in all the novels I have read as part of the project. How can the performance of our humanity offer us consolation from the evolutionary nature of life?

In answering Darwinian evolution's paranoid quest for biological truth, and the implicitly paranoid thesis arising out of this which states that human life is ontologically unimportant, these novelists offer rich combinations of reparative as well as paranoid modes of relating Darwin's thought. The narratives of the novels I have discussed confirm the inescapable nature of hereditary fate and perform an acceptance of the conspiratorial thesis of biological determinism. On the other hand, we can read Zola's literary Naturalism as drawing attention to its own scientific posture and the performative nature of its own truth. In this way, Zola's fiction is a vessel through which we can direct our suspicion at the truth claims made by Darwinian science and heredity determinism. In *Germinal*, Zola rehearses opposing interpretations of Darwin's theory of evolution, both of which are themselves paranoid in nature: the anarchist, Souvarine, insists on the intrinsic truth of his reading that Darwinian natural selection abnegates the possibility of all revolutionary action; while, in response, Étienne demands the same status for his own historically determinist, Marxist reading of Darwin. This dialogue, in which each man seeks to reveal the false consciousness of the other's Darwinism, gives literary form to a paranoid attitude in relation to Darwin's theory of natural selection. Zola presents Darwin's work as a body of thought open to conflicting interpretations, implicitly dismantling the objectivity on which literary as well as scientific naturalism subsists.

Yet, Zola's deconstructive, literary performance in *Germinal* is also a reparative one. Staging Darwin's theory of evolution in terms of irreconcilable possibilities denies natural selection the insidious, various, and mutually reinforcing social iterations it creates. In this way, the novel opens an interval between discourse and life, offering humanity a space in which to constitute itself in relation to the discourses of evolutionary science without necessarily being overcome by them. If the dialogic combination of Souvarine's nihilism and Étienne's utopianism can be read as a form of resistance to an implacable scientific objectivity, the dual figures of Claude and Sandoz in *L'Œuvre* represent actively transformative attitudes towards the grim reality of evolutionary truth. Although their theories of art are implicitly involved in questioning the scientific theories with which it is engaged, they each aim to wrest some redemptive potential from that same world view. In his plan to create a revolutionary artwork that represents the scientific narrative of the creation of the world, Sandoz negotiates truth as it is constructed and experiments with the consequences of that construction, in the hope of transforming the conditions of human life. Claude attempts a reparative relation to death through its artistic representation and confrontation. The irony of this reparative aesthetic is that the artist's search for redemption is presented by Zola as false consciousness even as he performs it himself. Zola's revolutionaries are failures. The laws of biological struggle and hereditary determinism reassert themselves, and those seeking to transform the natural world become victims of the violence to which they sought to respond. But by bearing witness to that repeated failure, to humanity's struggle against its own violent condition, Zola offers us a violent and brutal vision of life that functions as a performative reassertion of humanity's recuperative, creative desire.

Thomas Hardy's literary responses to Darwin present us with their own brand of suspicion and restorative optimism. Hardy's attitude towards the truth of the Darwinian paranoia — the possibility that human life is contingent, cruel, and devoid of intrinsic value — is one of reparative acceptance. Less concerned with the truth or untruth of this thesis, Hardy's work instead explores how everyday human life lives under its shadow. His depictions of the experiences of everyday female eroticism and sexuality in *The Return of the Native* in particular evince a reparative relation to evolutionary destiny and its temporality. Focusing on the cultural virtuosity of sexual desire and its rituals Hardy refuses to measure life's value in relation to its inevitable death, and depicts instead its fertility and creativity.

In his depiction of the relations of the sexes, Hardy focuses on what Sedgwick calls 'pleasure and amelioration', on the affective realities of people seeking to live affirmatively in relation to their own evolutionary fate. Under a regime of paranoid reading, such a focus is vulnerable to accusations of being 'merely' reformist or aesthetic, rather than rigorous or universally applicable.[4] But this emphasis in Hardy on pleasure and aesthetics as evolutionary meliorism offers a critical attitude towards evolutionary temporality and the determinist logic that derives from it. Through my engagement with Grosz's work on sexual selection, I have sought to

4 Ibid., p. 144.

show how Hardy's rituals in *The Return of the Native* are not limited to a doomed resistance of evolutionary fatalism. In his depiction of the relation of contingent, unpredictable cultural life with natural life, Hardy shows how desire and art open up new creative, semiotic, and erotic possibilities outside the immediate narrative frame of the novel. In Sedgwick's terms his attitude to human life is opposed to the paranoid attitude which seeks to anticipate and eliminate the unexpected. In Hardy's materialist and naturalist literary cosmology, aesthetic surprise is inherent to evolutionary contingency, re-emphasising how in Darwin's own work the paranoid neurosis of natural selection is supplemented by the descriptive, reparative work of sexual selection. *On the Origin of Species* is dedicated to unfolding the thesis that the conditions of natural life consist of random, contingent processes of extinction and variation. By contrast, *The Descent of Man* can be read as a descriptive text which details the myriad ways in which life elaborates itself, extends itself, and reproduces itself unpredictably. In this way, Hardy's fiction rebukes the attempt of evolutionary science to arm itself against surprise and offers us a vision of a materialism that stands in readiness, potentially at least, to appreciate the new.

Like Zola, Hardy's fiction is a testimony to the epistemological consequences of a paranoid response to Darwin's original thesis. His fiction, I have tried to show, details with unflinching candour the fatally sexist biological determinism of Darwin's theory of sexual difference as it becomes naturalised in Victorian society. But where in Zola the act of witnessing can itself be viewed as reparative, Hardy supplements this reparation further with a reading of Darwin's theory of sexual selection which allows him to recast of the type of agency available to humans in the evolutionary world. Hardy eschews the breathless, messianic hope of Zola's revolutionaries, but also the withdrawal from life involved in the act of covert observation. In Hardy's literary accounts, the rituals of sexual selection, evolutionary eroticism, and aesthetic participation offer us a reparative partaking in natural life without becoming engulfed by it.

Butler's, Huxley's, and Houellebecq's Utopian responses to Darwin give dialectical form to the relationship between suspicious modes and restorative modes of reading Darwin's theory of evolution. On the one hand, the spatially and semantically closed figure of the Utopian island represents the impossible, transcendent core of truth towards which a paranoid mode of reading travels. And through the construction of this truth, the Utopian island takes up a paranoid position in relation to the source material from which it was originally derived. As with More's *Utopia*, the Utopian islands of Butler, Huxley, and Houellebecq function as critiques of the contemporary historical, political, and biological conditions under which they written. A perfected, stable future can only be conceived in opposition to a degraded, unstable present. Above all, in these novels the sense that Darwin's theory of evolution is complicit in that degradation confirms the paranoid reading of Darwin: life is inescapably cruel and pointless.

On the other hand, these novels explore the difficulty that evolutionary incompletion presents to humanity's affective life and existential safety. In response, they affirm the reparative power of the Utopian imagination. Unlike Zola's and Hardy's works which give precedence to the act of bearing witness, for the protagonists in these Utopian novels the condition of the human under the regime of Darwinian evolution is too painful to even observe. Travelling to an existence that is temporally, geographically, and biologically displaced from the Darwinian world allows them to escape that pain momentarily. And for the reader, the imaginative act of reading these Utopian novels delivers us from the hegemony of the Darwinian conspiracy, reacquainting us with the possibility of imagining another reality. To make this reading a reparative act, paradoxically, requires that these Utopian futures be revealed as impossible. The destruction of Erewhon, Pala, and the neohuman in these novels respectively represents a rejection of the paranoiac epistemology. Acknowledging that evolutionary temporality entails the renewal as well as the destruction of all life is the starting point of an acceptance.

In such a light I have avoided making paranoid pronouncements which seek to fix the Darwinian context of these novels. It has been enough for me that these novelists were aware of Darwin's fundamental thesis and its scientific importance and could respond to that thesis in their work. It is also in the spirit of reparation that I have turned repeatedly to the work of Deleuze to try and read these novels' responses to Darwin in a positive light. Deleuze's reading of Darwin is unusual in the broad spectrum of critical theory and philosophy. Unlike many of his contemporary peers in critical theory, Deleuze's relation to biological discourse is not defined by suspicion or paranoia, but by a readiness to take what is useful for his own philosophical project without insisting on the transcendent truth of that project or the work he adapts for it. I have attempted to follow that example myself by exploring how literary authors enact this mode of reading Darwin's work, and how this enactment allows readers to appreciate the complexity, virtuosity, and philosophical ambition of these novelists' engagement with Darwin's theory of evolution.

But what of the paranoid dimension that remains of Darwin's work? As Sedgwick notes, paranoia is contagious. When suspicion is deployed as a primary methodology, she writes, it 'seems to grow like a crystal in a hypersaturated solution, blotting out any sense of possibility of alternative ways of understanding *or* things to understand'.[5] Such a remark in relation to the methodology of Darwin seems prescient when viewed in light of the increasingly reductive, essentialist, and unreflective iterations of Darwinian thought that have become prominent in the humanities as well as in specifically scientific contexts. As a so-called 'strong theory', the goal of Darwinian evolution in almost all its iterations seems to be reduced to as disparate a range of phenomena as possible to as few causal origins as possible. But this constant search for a mechanistic coherence in the natural world threatens to blot out the specific contingencies of life, the diversity of its forms, the creativity of its living, and its affective reality. For his own part, this constant drive to reduce difference to sameness exhausted Darwin. I have alluded before to the letter he writes to his wife from the hydrotherapeutic clinic at Moor Park where he was recovering from chronic stomach pains. Lying in the grass amidst the sound

5 Ibid., p. 131.

of birdsong and the skittering play of squirrels, he relinquishes for a moment the desire to discover the origins of all things: 'I did not care one penny how any of the beasts or birds had been formed'.[6] Deleuze notes that Darwin may have failed to see the value of his own work — the way in which it does not reduce difference to sameness but privileges the irreducible difference of all natural life and the constant action of difference in the natural world.

I have tried in my readings to articulate how these novelists' complex and dialectical responses to Darwin seek to rethink the nature and position of the human in the wake of evolution's Copernican revolution in the nineteenth century. Their works, like Darwin's, reassert the tragic nature of evolution: in these novels, life for humanity is grim, short, defined by relentless struggle and the impossibility of its escape. It is precisely in failing to articulate an image of human transcendence that these works offer us a sense of the reparative power of literature; of its capacity to respond to and reflect upon humanity's apparent demise; and of the necessary capacity to imagine a future for humanity, even as the struggle for life continues to rage.

6 Charles Darwin, letter to Emma Darwin, 28 April 1858, Darwin Correspondence Project <http://www.darwinproject.ac.uk/entry-2261> [accessed 3 November 2014].

BIBLIOGRAPHY

AMIGONI, DAVID, '"The Written Symbol Extends Infinitely": Samuel Butler and the Writing of Evolutionary Theory', in *Samuel Butler, Victorian against the Grain: A Critical Overview*, ed. by James G. Paradis (University of Toronto Press, 2007), pp. 91–112

ANSELL-PEARSON, KEITH, *Germinal Life: The Difference and Repetition of Deleuze* (Routledge, 1999)

——'Spectropoiesis and Rhizomatics: Learning to Live with Death and Demons', in *Evil Spirits: Nihilism and the Fate of Modernity*, ed. by Gary Banham and Charlie Blake (Manchester University Press, 2000), pp. 124–46

ARMSTRONG, NANCY, *Desire and Domestic Fiction: A Political History of the Novel* (Oxford University Press, 1987)

ASHER, DAVID, 'Schopenhauer and Darwinism', *Journal of Anthropology*, 1.3 (1871), pp. 312–32

ASQUITH, MARK, *Thomas Hardy, Metaphysics and Music* (Springer, 2005)

AUERBACH, ERICH, *Mimesis: The Representation of Reality in Western Literature*, trans. by Willard R. Trask (Princeton University Press, 2013)

BAGULEY, DAVID, 'Darwin, Zola and Dr Prosper Lucas's "Treatise on Natural Heredity"', in *The Literary and Cultural Reception of Charles Darwin in Europe*, ed. by Thomas F. Glick and Elinor Shaffer, 4 vols (Continuum, 2008–14), III, pp. 416–31

——*Naturalist Fiction: The Entropic Vision* (Cambridge University Press, 1990)

——'Zola and Darwin: A Reassessment', in *The Evolution of Literature: Legacies of Darwin in European Cultures*, ed. by Nicholas Saul and Simon J. James (Rodopi, 2011), pp. 201–12

BAILEY, J. O., 'Hardy's "Imbedded Fossil"', *Studies in Philology*, 42.3 (1945), pp. 663–74

BAKER, ROBERT S., 'Brave New World: Huxley's Dystopian Dilemma', in *Aldous Huxley's 'Brave New World'* (Chelsea House, 2003), pp. 93–104

BAKER-SMITH, DOMINIC, *More's Utopia* (University of Toronto Press, 2000)

BARAD, KAREN, *Meeting the Universe Halfway: Quantum Physics and the Entanglement of Matter and Meaning* (Duke University Press, 2007)

BARRETT, WILLIAM, 'Atlantic Bookshelf', *Atlantic Monthly*, April 1962, pp. 150–60

BATAILLE, GEORGES, *The Accursed Share: An Essay on General Economy. Volume 1: Consumption*, trans. by Robert Hurley (Zone Books, 1988)

——*Eroticism*, trans. by Mary Dalwood (Penguin Classics, 2012)

BEAUVOIR, SIMONE DE, *The Second Sex*, ed. by Sheila Malovany-Chevallier, trans. by Constance Borde (Vintage Books, 2011)

BECKER, COLETTE, 'Du meurtrier par hérédité au héros révolutionnaire: Étienne Lantier dans le dossier préparatoire de *Germinal*' <http://www.item.ens.fr/articles-en-ligne/> [accessed 19 July 2024]

——*Émile Zola, la fabrique de Germinal: dossier préparatoire de l'œuvre* (SEDES, 1986)

BEER, GILLIAN, 'Butler, Memory, and the Future', in *Samuel Butler, Victorian against the Grain: A Critical Overview*, ed. by James. G Paradis (University of Toronto Press, 2007), pp. 45–57

——*Darwin's Plots: Evolutionary Narrative in Darwin, George Eliot and Nineteenth-Century Fiction*, 3rd edn (Cambridge University Press, 2009)

BELSEY, CATHERINE, *Critical Practice*, 2nd edn (Routledge, 2003)

BENDER, BERT, *The Descent of Love: Darwin and the Theory of Sexual Selection in American Fiction, 1871–1926* (University of Pennsylvania Press, 1996)

BENJAMIN, WALTER, 'Theses on the Philosophy of History', in *Illuminations*, trans. by Harry Zohn (Houghton Mifflin Harcourt, 1968), pp. 253–64

BENNETT, JANE, *Vibrant Matter: A Political Ecology of Things* (Duke University Press, 2010)

BERG, WILLIAM J., *The Visual Novel: Emile Zola and the Art of His Times* (Pennsylvania State University Press, 1992)

BERG, WILLIAM J., and LAUREY KRAMER MARTIN, *Émile Zola Revisited* (Twayne, 1992)

BERNARDINI, JEAN-MARC, *Le Darwinisme social en France (1859–1918): fascination et rejet d'une idéologie* (CNRS, 1997)

BEST, VICTORIA, and MARTIN CROWLEY, *The New Pornographies: Explicit Sex in Recent French Fiction and Film* (Manchester University Press, 2007)

BETTY, LOUIS, *Without God: Michel Houellebecq and Materialist Horror* (Penn State University Press, 2016)

BIRNBAUM, MILTON, *Aldous Huxley: A Quest for Values* (Transaction, 2005)

BJÖRK, LENNART A., 'Annotation 980', in *The Literary Notebooks of Thomas Hardy*, ed. by Lennart A. Björk, 2 vols (New York University Press, 1985), I, p. 341

BLACKWELL, ANTOINETTE LOUISA BROWN, *The Sexes throughout Nature* (Hyperion Press, 1976)

BLATHWAYT, RAYMOND, 'A Chat with the Author of Tess', *Black and White*, 27 August 1892, p. 240

BLOCH, ERNST, *The Principle of Hope*, trans. by Neville Plaice, Stephen Plaice and Paul Knight, 3 vols (MIT Press, 1995)

BOOTH, WAYNE, 'Yes, But Are They Really Novels?', *Yale Review*, 51.4 (1962), pp. 630–36

BORDAS, ÉRIC, 'Ironie de l'ironie', in *Les Lieux du réalisme: pour Philippe Hamon*, ed. by Vincent Jouve and Alain Pagès (L'Improviste, 2005)

BOWERING, PETER, *Aldous Huxley: A Study of the Major Novels* (A. & C. Black, 2014)

BOWIE, MALCOLM, *Freud, Proust and Lacan: Theory as Fiction* (Cambridge University Press, 1987)

BOWLER, PETER J., *Evolution: The History of an Idea*, rev. edn (University of California Press, 1989)

BOYD, BRIAN, *On the Origin of Stories: Evolution, Cognition, and Fiction*, (Belknap Press of Harvard University Press, 2010)

BRAY, ABIGAIL, 'Not Woman Enough: Irigaray's Culture of Difference', *Feminist Theory*, 2.3 (2001), pp. 311–27, doi:10.1177/14647001010100200304

BRUNETIÈRE, FERDINAND, *L'Évolution des genres dans l'histoire de la littérature: leçons professées à l'École normale supérieure* (Hachette, 1890)

BUTLER, JUDITH, *Gender Trouble* (Routledge, 2002)

BUTLER, SAMUEL, 'Darwin among the Machines', in *A First Year in Canterbury Settlement with Other Early Essays*, ed. by R. A. Streatfeild (A. C. Fifield, 1914), pp. 180–85

—— 'The Deadlock in Darwinism', in *Essays on Life, Art, and Science*, ed. by R. A. Streatfeild (A. C. Fifield, 1908), pp. 234–340

—— *Erewhon*, ed. by Peter Mudford (Penguin, 1970)

—— 'Life and Habit, Vol. 2' in *The Shrewsbury Edition of the Works of Samuel Butler*, ed. by Henry Festing Jones and A. T. Bartholomew, 20 vols (Dutton, 1923–26), XVIII, pp. 111–84

—— 'Lucubratio Ebria', in *A First Year in Canterbury Settlement with Other Early Essays*, ed. by R. A. Streatfield (A. C. Fifield, 1914), pp. 186–95

CARROLL, JOSEPH, *Literary Darwinism: Evolution, Human Nature, and Literature* (Routledge, 2004)

——*Reading Human Nature: Literary Darwinism in Theory and Practice* (State University of New York Press, 2011)

CAYGILL, HOWARD, 'The Topology of Selection: The Limits of Deleuze's Biophilosophy', in *Deleuze and Philosophy: The Difference Engineer*, ed. by Keith Ansell-Pearson (Routledge, 1997), pp. 149–62

CECIL, DAVID, *Hardy the Novelist* (Constable, 1943)

CHAMBERS, ROBERT, *Vestiges of the Natural History of Creation* (John Churchill, 1844)

CHEAH, PHENG, and OTHERS, 'The Future of Sexual Difference: An Interview with Judith Butler and Drucilla Cornell', *Diacritics*, 28.1 (1998), pp. 19–42

COLATRELLA, CAROL, 'Feminism at War: Sexual Selection, Darwinism, and Fin-de Siècle Fiction', in *After Darwin: Literature, Theory, and Criticism in the Twenty-First Century*, ed. by Devin Griffiths and Deanna K. Kreisel (Cambridge University Press, 2023), pp. 108–20

COMTE, AUGUSTE, *Early Political Writings*, ed. by H. S. Jones (Cambridge University Press, 1998)

CROSBY, ALFRED W., *Ecological Imperialism: The Biological Expansion of Europe, 900–1900*, 2nd edn (Cambridge University Press, 2004)

DARWIN, CHARLES, *The Complete Work of Charles Darwin Online*, ed. by John van Whye <http://darwin-online.org.uk/> [accessed 15 May 2025]

——*The Descent of Man, and Selection in Relation to Sex*, 2 vols (John Murray, 1871) <https://darwin-online.org.uk/content/frameset?itemID=F937.1&viewtype=text&pageseq=1> [accessed 15 May 2025]

——*On the Origin of Species by Means of Natural Selection, or the Preservation of Favoured Races in the Struggle for Life*, 2nd edn, 2 vols (John Murray, 1860) <https://darwin-online.org.uk/content/frameset?itemID=F376&viewtype=text&pageseq=1> [accessed 15 May 2025]

——*De l'origine des espèces par sélection naturelle, ou, Des lois de transformation des êtres organisés*, trans. by Clémence Royer (Guillaumin, 1870)

——*Narrative of the Surveying Voyages of His Majesty's Ships Adventure and Beagle between the Years 1826 and 1836, Describing Their Examination of the Southern Shores of South America, and the Beagle's Circumnavigation of the Globe*, ed. by Robert Fitzroy, 3 vols (Henry Colburn, 1839)

DAWSON, WILLIAM J., 'Thomas Hardy', in *The Makers of English Fiction* (Fleming H. Revell, 1905), pp. 213–40

DEESE, R. S., *We Are Amphibians: Julian and Aldous Huxley on the Future of Our Species* (University of California Press, 2014)

DELANDA, MANUEL, *Intensive Science and Virtual Philosophy* (Bloomsbury, 2013)

DELEUZE, GILLES, *Desert Islands and Other Texts, 1953–1974*, ed. by David Lapoujade, trans. by Michael Taormina (Semiotext(e)/ MIT Press, 2004)

——*Difference and Repetition*, trans. by Paul Patton (Continuum, 2004)

——*The Logic of Sense*, ed. by Mark Lester, trans. by Constantin V. Boundas and Charles Stivale (Columbia University Press, 1990)

DELEUZE, GILLES, and FÉLIX GUATTARI, *Anti-Oedipus: Capitalism and Schizophrenia*, trans. by Robert Hurley, Mark Seem and Helen R. Lane (University of Minnesota Press, 1983)

DENNETT, DANIEL C., *Darwin's Dangerous Idea: Evolution and the Meanings of Life* (Penguin, 1996)

DENTITH, SIMON, 'Imagination and Inversion in Nineteenth-Century Utopian Writing', in *Anticipations: Essays on Early Science Fiction and Its Precursors*, ed. by David Seed (Syracuse University Press, 1995), pp. 137–52

DERRIDA, JACQUES, *The Animal that Therefore I Am*, trans. by David Wills (Fordham University Press, 2008)

——*Of Grammatology*, trans. by Gayatri Chakravorty Spivak (Johns Hopkins University Press, 1998)

DIMOCK, WAI CHEE, 'The Survival of the Unfit', in *After Darwin: Literature, Theory, and Criticism in the Twenty-First Century*, ed. by Devin Griffiths and Deanna K. Kreisel (Cambridge University Press, 2023), pp. 121–37

DISALVO, DAVID, 'What Is Literary Darwinism? An Interview with Joseph Carroll', *Neuronarrative*, 27 February 2009 <https://neuronarrative.wordpress.com/2009/02/27/what-is-literary-darwinism-an-interview-with-joseph-carroll/> [accessed 1 May 2016]

DUNAWAY, DAVID KING, *Huxley in Hollywood* (Harper & Row, 1989)

EBBATSON, ROGER, *The Evolutionary Self: Hardy, Forster, Lawrence* (Rowman & Littlefield, 1982)

ELDREDGE, NILES, *Darwin: Discovering the Tree of Life* (Norton, 2005)

ELLIOTT, ROBERT C., *The Shape of Utopia: Studies in a Literary Genre* [1970], ed. by Phillip E. Wegner (Peter Lang, 2013)

ENGELS, EVE-MARIE, *The Reception of Charles Darwin in Europe* (Continuum, 2008)

GLICK, THOMAS F., and ELINOR SHAFFER (eds), *The Literary and Cultural Reception of Charles Darwin in Europe*, 4 vols (Continuum 2008–14)

ENGELS, FRIEDRICH, 'The Part Played by Labour in the Transition from Ape to Man' in *Dialectics of Nature*, trans. by Clemens Dutt, 2nd rev. edn (Progress, 1986), pp. 170–83

ENGLAND, RICHARD, 'Censoring Huxley and Wilberforce: A New Source for the Meeting that the Athenaeum "Wisely Softened Down"', *Notes and Records of the Royal Society of London*, 71.4 (2017), pp. 371–84, doi:10.1098/rsnr.2016.058

FICHMAN, MARTIN, *Evolutionary Theory and Victorian Culture* (Humanity Books, 2002)

FIROR, RUTH, *Folkways in Thomas Hardy* (University of Pennsylvania Press, 1931)

FREEMAN, RICHARD BROKE, *The Works of Charles Darwin: An Annotated Bibliographical Handlist*, 2nd edn (Archon, 1977)

FREUD, SIGMUND, *Beyond the Pleasure Principle*, in *The Standard Edition of the Complete Psychological Works of Sigmund Freud*, ed. & trans. by James Strachey and others, 24 vols (Hogarth Press/ Institute of Psycho-Analysis, 1953–74), XVIII (1955), pp. 7–64

——*Civilization and Its Discontents*, in *The Standard Edition of the Complete Psychological Works of Sigmund Freud*, ed. & trans. by James Strachey and others, 24 vols (Hogarth Press/ Institute of Psycho-analysis, 1953–74), XXI (1961), pp. 57–146

——*Introductory Lectures on Psycho-Analysis (1916–1917)*, in *The Standard Edition of the Complete Psychological Works of Sigmund Freud*, ed. & trans. by James Strachey and others, 24 vols (Hogarth Press/ Institute of Psycho-Analysis, 1953–74), XVI (1963)

GALTON, FRANCIS, *Hereditary Genius: An Inquiry into Its Laws and Consequences*, 2nd edn (Macmillan, 1892)

GARWOOD, HELEN, *Thomas Hardy: An Illustration of the Philosophy of Schopenhauer* (John C. Winston, 1911)

GILLOTT, DAVID, *Samuel Butler against the Professionals: Rethinking Lamarckism 1860–1900* (Legenda, 2015)

GIORDANO, FRANK R., 'Eustacia Vye's Suicide', *Texas Studies in Literature and Language*, 22.4 (1980), pp. 504–21

GODWIN, WILLIAM, *An Enquiry Concerning Political Justice, and Its Influence on General Virtue and Happiness*, 2 vols (Robinson, 1793)

GONCOURT, EDMOND DE, and JULES DE GONCOURT, *Journal des Goncourts: mémoires de la vie littéraire*, 3 vols (Charpentier, 1887–88)

GOOCH, JOSHUA A., 'Figures of Nineteenth-Century Biopower in Samuel Butler's *Erewhon*', *Nineteenth-Century Contexts*, 36.1 (2014), 53–71 (pp. 54–57), doi:10.1080/08905495.2013.867599

GORDON, RAE BETH, *Dances with Darwin, 1875–1910: Vernacular Modernity in France* (Ashgate, 2009)

GOWATY, PATRICIA, 'Introduction: Darwinian Feminists and Feminist Evolutionists', in *Feminism and Evolutionary Biology: Boundaries, Intersections, and Frontiers*, ed. by Patricia Gowaty (Chapman & Hall, 1997), pp. 1–18

GRAFF, ANN-BARBARA, '"Administrative Nihilism": Evolution, Ethics and Victorian Utopian Satire', *Utopian Studies*, 12.2 (2001), pp. 33–52

GRIFFITHS, DEVIN, 'Darwin and Literature', in *The Cambridge Companion to Literature and Science*, ed. by Steven Meyer (Cambridge University Press, 2018), pp. 62–80

GRIFFITHS, DEVIN, and DEANNA K. KREISEL, 'After Darwin: Ecology, Posthumanism, and Aesthetics in the Twenty-First Century', in *After Darwin: Literature, Theory, and Criticism in the Twenty-First Century*, ed. by Devin Griffiths and Deanna K. Kreisel (Cambridge University Press, 2023), pp. 1–19

GRJEBINE, LIV, 'A Darwinian Murder: The Role of the Barré-Lebiez Affair in the Diffusion of Darwinism in Nineteenth-Century France', *Isis*, 113.4 (2022), pp. 689–709, doi:10.1086/722539

GROSZ, ELIZABETH, *Becoming Undone: Darwinian Reflections on Life, Politics, and Art* (Duke University Press, 2011)

—— 'Darwin and Feminism: Preliminary Investigations for a Possible Alliance', *Australian Feminist Studies*, 14.29 (1999), pp. 31–45, doi:10.1080/08164649993317

—— 'The Nature of Sexual Difference: Irigaray and Darwin', in *Sexual Difference Between Psychoanalysis and Vitalism*, ed. by Arun Saldanha and Hoon Song (Routledge, 2014), pp. 69–93

—— *The Nick of Time: Politics, Evolution, and the Untimely* (Duke University Press, 2004)

H., J., 'Schopenhauer-Darwin: Pessimismus Oder Optimismus', *Nature*, 85.2152 (1911), p. 403, doi:10.1038/085403a0

HALE, PIERS J., *Political Descent: Malthus, Mutualism, and the Politics of Evolution in Victorian England* (University of Chicago Press, 2014)

HALLWARD, PETER, *Out of This World: Deleuze and the Philosophy of Creation* (Verso, 2006)

HALPERIN, JOHN, 'Leslie Stephen, Thomas Hardy, and "A Pair of Blue Eyes"', *Modern Language Review*, 75.4 (1980), pp. 738–45, doi:10.2307/3726582

HARDY, FLORENCE EMILY, *The Early Life of Thomas Hardy, 1840–1891* (Cambridge University Press, 2011)

HARDY, THOMAS, 'Candour in English Fiction', in *Thomas Hardy's Public Voice: The Essays, Speeches, and Miscellaneous Prose*, ed. by Michael Millgate (Clarendon Press, 2001), pp. 95–102

—— *Late Lyrics and Earlier* (Macmillan & Co, 1922)

—— *The Literary Notebooks of Thomas Hardy*, ed. by Lennart A. Björk, 2 vols (New York University Press, 1985) <http://archive.org/details/literarynotebook0001hard> [accessed 19 May 2025]

—— *A Pair of Blue Eyes*, ed. by Alan Manford (Oxford University Press, 2005)

—— *The Return of the Native*, ed. by Simon Avery (Broadview Press, 2013)

—— 'The Science of Fiction', *The New Review*, 4 (1891), pp. 315–19; repr. in *Thomas Hardy's Public Voice: The Essays, Speeches, and Miscellaneous Prose*, ed. by Michael Millgate (Oxford University Press, 2001), pp. 106–10

HARAWAY, DONNA, *The Companion Species Manifesto: Dogs, People and Significant Otherness* (University of Chicago Press, 2003)

—— 'A Cyborg Manifesto: Science, Technology, and Socialist-Feminism in the Late Twentieth Century', in *Simians, Cyborgs and Women: The Reinvention of Nature* (Routledge, 1991), pp. 149–81

Harrow, Susan, *Zola, the Body Modern: Pressures and Prospects of Representation* (Legenda, 2010)

Harvey, David, 'Globalization and the "Spatial Fix"', *Geographische Revue: Zeitschrift für Literatur und Diskussion*, 3.2 (2001), pp. 23–30

Harvey, Joy Dorothy, *Almost a Man of Genius: Clémence Royer, Feminism, and Nineteenth-Century Science* (Rutgers University Press, 1997)

Hawkins, Mike, *Social Darwinism in European and American Thought, 1860–1945: Nature as Model and Nature as Threat* (Cambridge University Press, 1997)

Heiserman, A. R., 'Satire in the Utopia', *PMLA*, 78.3 (1963), pp. 163–74, doi:10.2307/460858

Hewitt, Martin, 'In the Wake of the Origin', in *The Reception of Darwinian Evolution in Britain, 1859–1909: Darwinism's Generations*, ed. by Martin Hewitt (Oxford University Press, 2024), pp. 88–163

Hillis Miller, J., *Thomas Hardy: Distance and Desire* (Belknap Press of Harvard University Press, 1970)

Hird, Myra J., 'Digesting Difference: Metabolism and the Question of Sexual Difference', *Configurations*, 20.3 (2012), pp. 213–37

Hobbes, Thomas, *Leviathan*, ed. by Richard Tuck (Cambridge University Press, 1991)

Hoffman, Anne Golomb, 'The Child in the Adult: Narrative and *Nachträglichkeit* in Henry James and Freud', *Psychoanalytic Quarterly*, 94.1 (2025), pp. 93–122, doi:10.1080/00332828.2024.2442117

Hofstadter, Richard, *Social Darwinism in American Thought*, rev. edn (Beacon Press, 1955)

Houellebecq, Michel, *H. P. Lovecraft: Against the World, against Life*, trans. by Dorna Khazeni (Gollancz, 2008)

——*Les Particules élémentaires* (Flammarion, 1998)

——*Atomised*, trans. by Frank Wynne (Vintage, 2001)

——*La Possibilité d'une île* [2005] (Fayard/ Livre de Poche, 2007)

——*The Possibility of an Island*, trans. by Gavin Bowd (Vintage, 2007)

Hubbard, Ruth, *The Politics of Women's Biology* (Rutgers University Press, 1990)

Huxley, Aldous, *Island*, ed. by David Bradshaw (Vintage Books, 2005)

Huxley, Julian, 'Charles Darwin: Galápagos and after', in *The Galápagos: Proceedings of the Symposia of the Galápagos International Scientific Project*, ed. by Robert I. Bowman (University of California Press, 1966), pp. 3–9

Irigaray, Luce, *An Ethics of Sexual Difference*, trans. by Carolyn Burke and Caroline C. Gill (Cornell University Press, 1993)

——*I Love to You: Sketch of A Possible Felicity in History*, trans. by Alison Martin (Routledge, 1996)

Irvine, William, 'The Influence of Darwin on Literature', *Proceedings of the American Philosophical Society*, 103.5 (1959), pp. 616–28

James, Henry, *The Art of Criticism: Henry James on the Theory and Practice of Criticism*, ed. by William Veeder and Susan M. Griffin (Chicago University Press, 1986)

Jameson, Fredric, *Archaeologies of the Future: The Desire Called Utopia and Other Science Fictions* (Verso, 2007)

Jeffery, Ben, *Anti-Matter: Michel Houellebecq and Depressive Realism* (Zero Books, 2011)

Johnson, Lionel, *The Art of Thomas Hardy* (Mathews & Lane, 1894)

Josipovici, Gabriel, *The World and the Book: A Study of Modern Fiction* (Stanford University Press, 1971)

Kaye, Richard A., *The Flirt's Tragedy: Desire without End in Victorian and Edwardian Fiction* (University of Virginia Press, 2002)

Kendrick, Christopher, 'More's Utopia and Uneven Development', *Boundary 2*, 13.2–3 (1985), pp. 233–66

KERMODE, FRANK, '"Island" by Aldous Huxley', *The Partisan Review*, 29.3 (1962), pp. 472–73

KIMBALL, M. DOUGLAS, 'Emile Zola and French Impressionism', *Bulletin of the Rocky Mountain Modern Language Association*, 23.2 (1969), pp. 51–57, doi:10.2307/1346694

KRISTEVA, JULIA, 'Aimer la vérité cruelle et disgracieuse', *Cahiers Naturalistes*, 68 (1994), pp. 7–9

——*Powers of Horror: An Essay on Abjection*, trans. by Leon S. Roudiez (Columbia University Press, 1982)

KUHN, THOMAS S., *The Structure of Scientific Revolutions*, 3rd edn (University of Chicago Press, 1970)

KUMAR, KRISHAN, *Utopianism* (University of Minnesota Press, 1991)

LACAN, JACQUES, *The Seminar of Jacques Lacan, Book XVII: The Other Side of Psychoanalysis*, trans. by Russell Grigg (Norton, 2007)

——'The Subversion of the Subject and the Dialectic of Desire in the Freudian Unconscious', in *Écrits: The First Complete Edition in English*, trans. by Bruce Fink (Norton, 2006), pp. 671–702

LAMARCK, JEAN BAPTISTE, *Zoological Philosophy: An Exposition with Regard to the Natural History of Animals*, trans. by Hugh Elliot Macmillan (University of Chicago Press, 1984)

LANDRY, TRAVIS, *Subversive Seduction: Darwin, Sexual Selection, and the Spanish Novel* (University of Washington Press, 2013)

LANOUX, ARMAND, *Bonjour, Monsieur Zola* (Hachette, 1962)

LARSON, EDWARD J., *Evolution's Workshop: God and Science on the Galápagos Islands* (Basic Books, 2001)

LAWRENCE, D. H., 'Study of Thomas Hardy', in *Study of Thomas Hardy and Other Essays*, ed. by Bruce Steele (Cambridge University Press, 1985), pp. 3–132

LEMAITRE, CLÉMENT, 'Mutations de l'ironie dans l'œuvre de Michel Houellebecq', *Carnets: Revue électronique d'études françaises de l'APEF*, 2nd ser., 23 (2022), doi:10.4000/carnets.13730

LEPS, MARIE-CHRISTINE, *Apprehending the Criminal: The Production of Deviance in Nineteenth Century Discourse* (Duke University Press, [1992])

LEVINE, GEORGE, *Darwin and the Novelists: Patterns of Science in Victorian Fiction* (Harvard University Press, 1988)

——*Darwin Loves You: Natural Selection and the Re-Enchantment of the World* (Princeton University Press, 2008)

——*Dying to Know: Scientific Epistemology and Narrative in Victorian England* (University of Chicago Press, 2002)

——'Foreword', in Gillian Beer, *Darwin's Plots: Evolutionary Narrative in Darwin, George Eliot and Nineteenth-Century Fiction*, 3rd edn (Cambridge University Press, 2009), pp. ix–xiv

LODGE, DAVID, '"The Woodlanders": A Darwinian Pastoral Elegy', in *Working with Structuralism: Essays and Reviews on Nineteenth- and Twentieth-Century Literature* (Routledge, 1981), pp. 79–94

LOGAN, GEORGE M., and ROBERT M. ADAMS, 'Introduction', in Thomas More, *Utopia*, ed. by George M. Logan and Robert M. Adams, trans. by Robert M. Adams (Cambridge University Press, 1989), pp. xi–xxix

LÓPEZ-BELTRÁN, CARLOS, 'In the Cradle of Heredity: French Physicians and *L'Hérédité Naturelle* in the Early 19th Century', *Journal of the History of Biology*, 37.1 (2004), pp. 39–72, doi:10.1023/b:hist.0000020385.44164.e2

LOVEJOY, ARTHUR O., 'Schopenhauer as an Evolutionist', *The Monist*, 21.2 (1911), pp. 195–222

LUCAS, J. R., 'Wilberforce and Huxley: A Legendary Encounter', *The Historical Journal*, 22.2 (1979), pp. 313–30

LUCAS, PROSPER, *Traité philosophique et physiologique de l'hérédité naturelle dans les états de santé et de maladie du système nerveux* (Baillière, 1847)

LYLE, LOUISE, '"Le Struggleforlife": Contesting Balzac through Darwin in Zola, Bourget, and Barrès', *Nineteenth-Century French Studies*, 36.3–4 (2008), pp. 305–19

MALTHUS, T. R., *An Essay on the Principle of Population, or, A View of Its Past and Present Effects on Human Happiness with an Inquiry into Our Prospects Respecting the Future Removal or Mitigation of the Evils Which It Occasions* (R. C. Weightman, 1809)

MARIN, LOUIS, *Utopics: The Semiological Play of Textual Spaces*, trans. by Robert Vollrath (Humanities Press International, 1990)

MARX, KARL, *Capital. Volume One: A Critique of Political Economy*, ed. by Friedrich Engels, trans. by Samuel Moore and Edward Aveling (Dover, 2012)

—— *Economic and Philosophic Manuscripts of 1844*, ed. and trans. by Martin Milligan (Dover, 2012)

MCCANN, JOHN, *Michel Houellebecq: Author of Our Times* (Peter Lang, 2010)

MCDOWALL, ARTHUR SYDNEY, *Thomas Hardy: A Critical Study* (Faber & Faber, 1931)

MENNINGHAUS, WINFRIED, *Disgust: The Theory and History of a Strong Sensation*, trans. by Howard Eiland and Joel Golb (State University of New York Press, 2003)

MITTERAND, HENRI, *Émile Zola: Fiction and Modernity* (Émile Zola Society, 2000)

MORE, THOMAS, *Utopia*, ed. by George M. Logan and Robert M. Adams, trans. by Robert M. Adams (Cambridge University Press, 1989)

MORREY, DOUGLAS, *Michel Houellebecq: Humanity and Its Aftermath* (Liverpool University Press, 2013)

MORTON, TIMOTHY, *Hyperobjects: Philosophy and Ecology after the End of the World* (University of Minnesota Press, 2013)

MUMFORD, LEWIS, 'Introduction', in Samuel Butler, *Erewhon, and Erewhon Revisited* (Modern Library, 1927), pp. xix–xxvii

NELSON, BRIAN, 'Zola and the Ideology of Messianism', *Orbis Litterarum*, 37.1 (1982), pp. 70–82, doi:10.1111/j.1600–0730.1982.tb00790.x

NOGUERA SOLANO, RICARDO, 'Darwin and Inheritance: The Influence of Prosper Lucas', *Journal of the History of Biology*, 42.4 (2009), pp. 685–714, doi:10.1007/s10739–008–9175–7

ORTEGA Y GASSET, JOSÉ, *Meditations on Quixote* (University of Illinois Press, 2000)

PARADIS, JAMES G., 'The Butler-Darwin Biographical Controversy in the Victorian Periodical Press', in *Science Serialized: Representations of the Sciences in Nineteenth-Century Periodicals*, ed. by Geoffrey Cantor and Sally Shuttleworth (MIT Press, 2004), pp. 307–29

—— 'Introduction', in *Samuel Butler, Victorian Against the Grain: A Critical Overview*, ed. by James G. Paradis (University of Toronto Press, 2007), pp. 1–18

PARRINDER, PATRICK, 'Eugenics and Utopia: Sexual Selection from Galton to Morris', *Utopian Studies*, 8.2 (1997), pp. 1–12

—— *Utopian Literature and Science: From the Scientific Revolution to Brave New World and Beyond* (Palgrave Macmillan, 2015)

PRUM, MICHEL, 'Charles Darwin's First French Translations', in *The Literary and Cultural Reception of Charles Darwin in Europe*, ed. by Thomas F. Glick and Elinor Shaffer, 4 vols (Bloomsbury, 2014), IV, pp. 391–99

RABBOSEAU, SANDRINE, 'Houellebecq ou le renouveau du roman expérimental', in *Michel Houellebecq sous la loupe*, ed. by Murielle Lucie Clément and Sabine van Wesemael (Rodopi, 2007), pp. 43–51

RAY, NICHOLAS, 'Interrogating the Human/Animal Relation in Freud's *Civilisation and Its Discontents*', *Humanimalia*, 6.1 (2014), pp. 10–40, doi:10.52537/humanimalia.9926

RESHE, JULIE, 'The Death Drive of Evolution (From the Perspective of Depressive Realism)', *Stasis*, 11.1 (2021), doi:10.33280/2310–3817–21–11–1–156–180

RICHARDS, JANET RADCLIFFE, *Human Nature after Darwin: A Philosophical Introduction* (Routledge, 2005)
RICHARDSON, ANGELIQUE, 'Hardy and Biology', in *Thomas Hardy: Texts and Contexts*, ed. by Phillip Mallet (Palgrave Macmillan, 2002), pp. 156–79
—— '"How I Mismated Myself for Love of You!": The Biologization of Romance in Hardy's "A Group of Noble Dames"', *Thomas Hardy Journal*, 14.2 (1998), pp. 59–76
—— '"Some Science Underlies All Art": The Dramatization of Sexual Selection and Racial Biology in Thomas Hardy's *A Pair of Blue Eyes* and *The Well-Beloved*', *Journal of Victorian Culture*, 3.2 (1998), pp. 302–38, doi:10.1080/13555509809505967
RICHTER, VIRGINIA, *Literature after Darwin: Human Beasts in Western Fiction 1859–1939* (Palgrave Macmillan, 2011)
RICŒUR, PAUL, *Freud and Philosophy: An Essay on Interpretation*, trans. by Denis Savage (Yale University Press, 1970)
RIMMER, MARY, 'Club Laws: Chess and the Construction of Gender in "A Pair of Blue Eyes"', in *The Sense of Sex: Feminist Perspectives on Thomas Hardy*, ed. by Margaret Higonnet (University of Illinois Press, 1993), pp. 203–19
ROSSER, SUE, *Biology and Feminism: A Dynamic Interaction* (Twayne, 1992)
ROYER, CLÉMENCE, 'Préface de la première édition', in Charles Darwin, *De l'origine des espèces par sélection naturelle, ou, Des lois de transformation des êtres organisés*, ed. and trans. by Clémence Royer, 3rd edn (Guillaumin, 1870), pp. xxvii–lxxi
RYAN, VANESSA L., 'Living in Duplicate: Victorian Science and Literature Today', *Critical Inquiry*, 38.2 (2012), pp. 411–17, doi:10.1086/662752
RYAZANSKAYA, S. W. (ed.), *Karl Marx and Friedrich Engels: Selected Correspondence* (Progress, 1965)
SAUL, NICHOLAS, and SIMON J. JAMES (eds), *The Evolution of Literature Legacies of Darwin in European Cultures* (Rodopi, 2011)
SCHAEFER, DONOVAN O., 'Darwin's Orchids: Evolution, Natural Law, and the Diversity of Desire', *GLQ: A Journal of Lesbian and Gay Studies*, 27.4 (2021), pp. 525–50, doi:10.1215/10642684–9316824
SCHOPENHAUER, ARTHUR, *The World as Will and Representation*, trans. by E. F. J. Payne, 2 vols (Dover, 1966–69)
SCHOR, NAOMI, *Zola's Crowds* (Johns Hopkins University Press, 1978)
SEDGWICK, EVE KOSOFSKY, 'Paranoid Reading, Reparative Reading, or You're So Paranoid, You Probably Think This Essay Is about You', in *Touching Feeling: Affect, Pedagogy, Performativity* (Duke University Press, 2003), pp. 123–51
SEEM, MARK, 'Introduction', in Gilles Deleuze and Félix Guattari, *Anti-Oedipus: Capitalism and Schizophrenia*, trans. by Robert Hurley, Mark Seem and Helen R. Lane (University of Minnesota Press, 1983), pp. xv–xxiv
SERRES, MICHEL, *Feux et signaux de brume: Zola* (Grasset, 1975)
SHIDELER, ROSS, *Questioning the Father: From Darwin to Zola, Ibsen, Strindberg, and Hardy* (Stanford University Press, 1999)
SHOSTAK, STANLEY, *Evolution of Sameness and Difference* (CRC Press, 1999)
SHUTTLEWORTH, SALLY, *George Eliot and Nineteenth-Century Science: The Make-Believe of a Beginning* (Cambridge University Press, 1984)
SION, RONALD T., *Aldous Huxley and the Search for Meaning: A Study of the Eleven Novels* (McFarland, 2010)
SNOW, C. P., *The Two Cultures* (Cambridge University Press, 2012)
SPENCER, HERBERT, 'From Liberty to Bondage (1891)', in *The Man versus the State: With Six Essays on Government, Society, and Freedom* (Liberty Classics, 1981), pp. 487–519
—— *The Principles of Biology* (William & Norgate, 1864)

SREENAN, NIALL, 'Dreaming of Islands: Individuality and Utopian Desire in Post-Darwinian Literature', *Island Studies Journal*, 12.2 (2017), pp. 267–80, doi:10.24043/isj.34
——'In the Name of the Father: Darwin, Scientific Authority and Literary Assimilation', in *From Shakespeare to Autofiction*, ed. by Martin Procházka (UCL Press, 2024), pp. 91–108
——'Universal, Acid: Houellebecq's Clones and the Evolution of Humanity', *Modern & Contemporary France*, 27.1 (2019), pp. 77–93, doi:10.1080/09639489.2018.1557130
STACK, DAVID [D. A.], 'Charles Darwin's Liberalism in "Natural Selection as Affecting Civilised Nations"', *History of Political Thought*, 33.3 (2012), pp. 525–54
——'The First Darwinian Left: Radical and Socialist Responses to Darwin, 1859–1914', *History of Political Thought*, 21.4 (2000), pp. 682–710
STACK, DAVID, 'Charles Darwin's Liberalism in "Natural Selection as Affecting Civilised Nations"', *History of Political Thought*, 33.3 (2012), pp. 525–54
STANNARD, DAVID E., *American Holocaust: The Conquest of the New World* (Oxford University Press, 1992)
STEINLIGHT, EMILY, 'Hardy's Unnecessary Lives: The Novel as Surplus', *Novel*, 47.2 (2014), pp. 224–41, doi:10.1215/00295132-2647167
STEVENSON, LIONEL, *Darwin among the Poets* (Russell & Russell, 1963)
STONE, ALISON, *Luce Irigaray and the Philosophy of Sexual Difference* (Cambridge University Press, 2006)
SUSSMAN, HERBERT L., *Victorian Technology: Invention, Innovation, and the Rise of the Machine* (Praeger, 2009)
SWEENEY, CAROLE, *Michel Houellebecq and the Literature of Despair* (A. & C. Black, 2013)
SYMONDS, JOHN ADDINGTON, 'On the Application of Evolutionary Principles to Art and Literature', in *Essays: Speculative and Suggestive*, 3rd edn (Smith, Elder, 1907), pp. 27–57
THOMAS, JANE, *Thomas Hardy and Desire: Conceptions of the Self* (Palgrave Macmillan, 2013)
THORNTON, HELEN, *State of Nature or Eden? Thomas Hobbes and His Contemporaries on the Natural Condition of Human Beings* (University of Rochester Press, 2005)
WATT, IAN, *The Rise of the Novel: Studies in Defoe, Richardson, and Fielding* (University of California Press, 1957)
WEBER, CARL JEFFERSON, *Hardy of Wessex: His Life and Literary Career* (Columbia University Press, 1940)
WEBER, MAX, *The Protestant Ethic and the Spirit of Capitalism*, trans. by Talcott Parsons (Routledge, 2010)
WEISMANN, AUGUST, *The Germ-Plasm: A Theory of Heredity*, trans. by W. Newton Parker and Harriet Rönnfeldt (Scribner, 1893)
WELLEK, RENÉ, and STEPHEN G. NICHOLS, *Concepts of Criticism* (Yale University Press, 1963)
WENG, GUSTAV, *Schopenhauer-Darwin: Pessimismus oder Optimismus? Ein Beitrag zur Fortschrittsbewegung* (Ernst Hofmann, 1911)
WILLIAMS, RAYMOND, 'Social Environment and Theatrical Environment', in *Culture and Materialism: Selected Essays* (Verso, 2005), pp. 88–104
WOLFE, CARY, *What Is Posthumanism?* (University of Minnesota Press, 2009)
WYNTER, SYLVIA, 'Unsettling the Coloniality of Being/ Power/ Truth/ Freedom: Towards the Human, after Man, Its Overrepresentation — An Argument', *CR: The New Centennial Review*, 3.3 (2003), pp. 257–337 (p. 308), doi:10.1353/ncr.2004.0015
YEAZELL, RUTH BERNARD, 'Nature's Courtship Plot in Darwin and Ellis', *Yale Journal of Criticism*, 2.2 (1989), pp. 33–53
YVOREL, JEAN-JACQUES, 'L'Enfant des Halles', in *Les Halles: images d'un quartier*, ed. by Jean-Louis Robert and Myriam Tsikounas (La Sorbonne, 2004), pp. 107–18
ZEMKA, SUE, '"Erewhon" and the End of Utopian Humanism', *ELH*, 69.2 (2002), pp. 439–72

ZOLA, ÉMILE, 'Différences entre Balzac et moi (1869)', in *Comment Émile Zola composait ses romans: d'après ses notes personnelles et inédites*, ed. by Henri Massé (Charpentier, 1906)
—— *Germinal*, ed. by Olivier Lumbroso and Henri Mitterand, in *Œuvres complètes. Tome 12: Souffrance et révolte (1884–1885)* (Nouveau Monde, 2005)
—— *Germinal*, trans. by Peter Collier (Oxford University Press, 2008)
—— *L'Œuvre*, ed. by Henri Mitterand and Olivier Got, in *Œuvres complètes. Tome 13: Naturalisme pas mort! (1886–1888)* (Nouveau Monde, 2005)
—— *The Masterpiece*, trans. by Roger Pearson and Thomas Walton, rev. edn (Oxford University Press, 2008)
—— *Le Roman expérimental*, 2nd edn (Charpentier, 1880)
—— *The Experimental Novel, and Other Essays*, trans. by Belle M. Sherman (Cassell, 1893)
—— *Le Ventre de Paris*, ed. by Henri Mitterand, Patricia Carles, and Béatrice Desgranges, in *Œuvres complètes. Tome 5: Thiers au pouvoir (1871–1873)* (Nouveau Monde, 2005)
—— *The Belly of Paris*, trans. by Brian Nelson (Oxford University Press, 2009)

INDEX

❖

Literary works are accessible from author entries. Abbreviations are expanded on page xi. Footnotes are given in the form '38 n. 15' (note 15 on page 38). ★ indicates fictional characters and places.

www.ingramcontent.com/pod-product-compliance
Lightning Source LLC
LaVergne TN
LVHW081933200926
848484LV00009B/2420

* 9 7 8 1 7 8 1 8 8 8 4 7 6 *